The Importance of
Scientific
Theory

The Importance of
Atomic Theory

John Allen

ReferencePoint
Press®

San Diego, CA

About the Author

John Allen is a writer living in Oklahoma City.

© 2016 ReferencePoint Press, Inc.
Printed in the United States

For more information, contact:
ReferencePoint Press, Inc.
PO Box 27779
San Diego, CA 92198
www.ReferencePointPress.com

Picture Credits:
Cover: Thinkstock Images; Maury Aaseng: 35; © Bettmann/Corbis: 41; BSIP, Laurant, Maya/Science Photo Library: 57; © Corbis: 45; David Nicholls/Science Photo Library: 38; Polaris/Newscom: 9; © Louie Psihoyos/ Corbis: 69; © Science Picture Co./Corbis: 61; Thinkstock Images: 6, 7, 14; © Carol & Mike Werner/Visuals Unlimited/Corbis: 22; Zephyr/Science Photo Library: 55; John Dalton, founder of chemical atomic theory, Jackson, Peter (1922–2003)/Private Collection/© Look and Learn/Bridgeman Images: 19; McGill University, Rutherford Museum/Emilio Segre Visual Archives/American Institute of Physics/Science Photo Library: 27; Albert Einstein: The Greatest Thinker of His Time. Einstein worked at numerous universities and polytechnics. Scan of small illustration which has been digitally enhanced to assist repro.: 31; Rontgen, Wilhelm Conrad (1845–1923). German physicist./Photo © Tarker/Bridgeman Images: 50

LIBRARY OF CONGRESS CATALOGING-IN-PUBLICATION DATA

Allen, John, 1957-
The Importance of Atomic Theory/by John Allen.
    pages cm.--(Importance of scientific theory)
Includes bibliographical references and index.
Audience: Grade 9 to 12.
ISBN-13: 978-1-60152-786-8 (hardback)
ISBN-10: 1-60152-786-1 (hardback)
1. Atomic theory--History. 2. Atomic theory--History--Study and teaching (Secondary)
3. Atomic theory--Study and teaching (Secondary) I. Title.
QD461.A394 2015
539'.14--dc23

                                                                    2014036911

CONTENTS

FOREWORD

What is the nature of science? The authors of "Understanding the Scientific Enterprise: The Nature of Science in the Next Generation Science Standards," answer that question this way: "Science is a way of explaining the natural world. In common parlance, science is both a set of practices and the historical accumulation of knowledge. An essential part of science education is learning science and engineering practices and developing knowledge of the concepts that are foundational to science disciplines. Further, students should develop an understanding of the enterprise of science as a whole—the wondering, investigating, questioning, data collecting and analyzing."

Examples from history offer a valuable way to explore the nature of science and understand the core ideas and concepts around which all life revolves. When English chemist John Dalton formulated a theory in 1803 that all matter consists of small, indivisible particles called atoms and that atoms of different elements have different properties, he was building on the ideas of earlier scientists as well as relying on his own experimentation, observation, and analysis. His atomic theory, which also proposed that atoms cannot be created or destroyed, was not entirely accurate, yet his ideas are remarkably close to the modern understanding of atoms. Intrigued by his findings, other scientists continued to test and build on Dalton's ideas until eventually—a century later—actual proof of the atom's existence emerged.

The story of these discoveries and what grew from them is presented in *The Importance of Atomic Theory*, one volume in ReferencePoint's series *The Importance of Scientific Theory*. The series strives to help students develop a broader and deeper understanding of the nature of science by examining notable ideas and events in the history of science. Books in the series focus on the development and outcomes of atomic theory, cell theory, germ theory, evolution theory, plate tectonic theory, and more. All books clearly state the core idea and explore changes in thinking over time, methods of experimentation and observation, and societal impacts of these momentous theories and discoveries. Each volume includes a visual chronology; brief descriptions of important people; sidebars that highlight and further explain key events and concepts; "words in context" vocabulary; and, where possible, the words of the scientists themselves.

Through richly detailed examples from history and clear discussion of scientific ideas and methods, *The Importance of Scientific Theory* series furthers an appreciation for the essence of science and the men and women who devote their lives to it. As the authors of "Understanding the Scientific Enterprise: The Nature of Science in the Next Generation Science Standards" write, "With the addition of historical examples, the nature of scientific explanations assumes a human face and is recognized as an ever-changing enterprise."

IMPORTANT DATES IN THE HISTORY OF ATOMIC THEORY

1680
Irish chemist Robert Boyle defines an element as something that cannot be broken down by chemical means, a perception that leads to the periodic table and ideas about atomic structure.

1900
German physicist Max Planck proposes the quantum theory, which says that electromagnetic energy can be emitted only in chunks called quanta (later renamed photons).

AD 1649
Pierre Gassendi, a French professor of mathematics, publishes a book that revives the idea that the universe is made up of atoms and the void.

1898
English physicist Joseph John Thomson discovers the electron and proposes that negatively charged electrons are scattered throughout the positively charged sphere of an atom—the plum-pudding atomic model.

BCE 600	1600	1700	1800	1900

1803
English chemist John Dalton develops an atomic theory that says all matter consists of small, indivisible particles called atoms; atoms cannot be created or destroyed; and atoms of different elements have different properties.

BCE ca. 400
The Greek philosopher Democritus develops the idea of his mentor Leucippus that the universe is made up of tiny, indivisible bits of matter called atoms.

1869
Russian chemist Dmitri Mendeleev creates the periodic table for a chemistry textbook, arranging elements according to atomic weights and properties.

1895
German physicist Wilhelm Roentgen discovers X-rays, which are charged electron particles flowing through a Crookes tube. The rays are able to pass through objects and produce images of the body's interior.

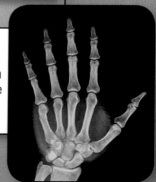

6

1911
New Zealander physicist Ernest Rutherford proposes that an atom is composed of a tiny, positively charged nucleus orbited by even tinier, negatively charged electrons. This is called the planetary atomic model.

2012
Scientists in Europe discover the Higgs boson, an elusive atomic particle that provides clues about the beginnings of the universe.

1923
Austrian physicist Erwin Schrödinger uses the idea of electrons moving in waves to explain Bohr's atomic theory. At the same time, German physicist Werner Heisenberg discovers a similar explanation using the idea of electrons as particles.

1927
Werner Heisenberg notes that it is impossible to know at the same time the location and velocity of an electron. This idea is called the Uncertainty Principle.

1960
American physicist Theodore Maiman builds the first working ruby-light laser.

1910   1930   1950   1970 / 2010

1913
Danish physicist Niels Bohr applies quantum theory to the atomic model. He describes how electrons emit quanta of energy as they move from higher energy levels to lower energy levels. Bohr also observes that chemical properties are determined by the number of electrons in the outer orbits of an atom.

1953
English scientists James Watson and Francis Crick describe the atomic structure of a DNA molecule as a three-dimensional double helix.

1945
A team of the world's top physicists produces the first atomic bomb for the US military's Manhattan Project.

1905
German physicist Albert Einstein publishes papers on Brownian motion and the particle nature of light. The former essentially proves the existence of atoms, and the latter leads to the idea of nuclear energy.

INTRODUCTION

A Theory of Tiny Particles with Large Impact

THE CORE IDEA

Atomic theory is a scientific theory about the nature of matter. It states that matter is composed of tiny units called atoms that are constantly in motion. Each atom consists of a core called the nucleus and surrounding particles called electrons. The nucleus contains two kinds of particles, called neutrons and protons. Atoms can be measured and compared by weight. Every element, such as carbon or oxygen, is made up entirely of one type of atom. The atoms of each element are different from the atoms of every other element. Atoms of one element can combine with those of another element to form molecules, which are chemical compounds. When joined into a molecule, the atoms of elements bond together in whole-number ratios that are specific and predictable. Atomic theory is the basis of a great deal of modern science, including physics and chemistry.

On July 4, 2012, scientists at a large facility on the French-Swiss border announced the results of a momentous experiment: the discovery of a new atomic particle called the Higgs boson. The particle, whose existence had been predicted but not verified, provides mass to other fundamental particles. It essentially makes the universe work the way it does. Physicists say the discovery is proof for the big bang theory of how the universe began. As English physicist Jon Butterworth describes the experiment: "Pretty much anything could in principle turn up, since no one did this before. But one thing—the Higgs boson—had to turn up, or our understanding of fundamental physics was incomplete. Well, let's be frank, wrong."[1] Scientists the world over were excited that modern physics was *not* wrong.

Public reaction however was muted. News reports trumpeted the discovery as the greatest of the twenty-first century, but ordinary people were mostly baffled. How scientists could make such sweeping claims on the basis of something so tiny and elusive seemed hard to credit. And so it has been with many of the breakthroughs related to atomic theory and the behavior of atoms and molecules. Perhaps no scientific theory affects daily life so profoundly, yet the average person understands it only vaguely, if at all. Discussing the Higgs boson find, American scientist Ainissa Ramirez says: "We have a small percentage of people [i.e., scientists] who are losing their minds, having huge parties and the rest of the

A computer simulation depicts the Higgs boson atomic particle, whose discovery was announced in July 2012. Physicists say the discovery is proof for the big bang theory of how the universe began.

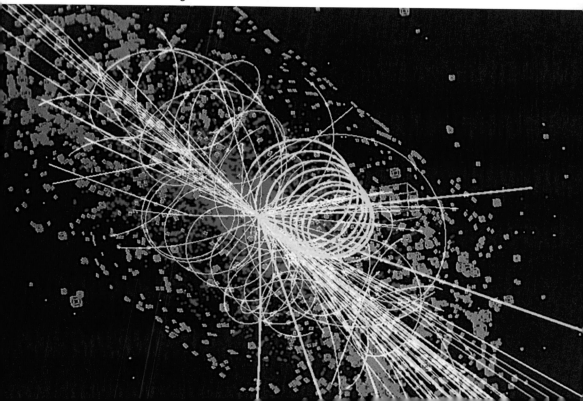

world doesn't really know what they're excited about. So if we spend a lot of money to put together these huge experiments—the Large Hadron Collider, where the Higgs boson was found, costs on the upwards of about $10 billion. Let's spend a little bit of that money just educating people."[2]

Atomic theory has led to everything from modern chemistry to nuclear energy. DVDs, MRIs, laser light shows, and thousands of other things people take for granted every day sprang from scientific discoveries about atoms and particles. The idea that the universe is made of tiny bits called atoms goes back to the ancient Greeks, and that perception is still fundamental today. As physicist Richard Feynman expressed it in a 1962 lecture:

> If, in some cataclysm, all of scientific knowledge were to be destroyed, and only one sentence passed on to the next generations of creatures, what statement would contain the most information in the fewest words? I believe it is the *atomic hypothesis* (or the atomic *fact*, or whatever you wish to call it) that *all things are made of atoms—little particles that move around in perpetual motion, attracting each other when they are a little distance apart, but repelling upon being squeezed into one another.* In that one sentence, you will see, there is an *enormous* amount of information about the world, if just a little imagination and thinking are applied.[3]

WORDS IN CONTEXT

subatomic
Relating to particles smaller than atoms.

Chemists in the last two centuries used insights about atomic structure to combine molecules and make all sorts of new materials. Researchers discovered the DNA molecule and mapped the genes of the entire human race. Scientists and engineers continue to use information about atoms to dream up the most amazing new technologies. Even the baffling behavior of electrons as described in quantum mechanics has been harnessed to make household products such as LED lights and laser-based measuring tools. Scientists predict that someday quantum computers will operate at speeds undreamt of today. However, atomic theory has also led to dangerous inventions that threaten human survival. The awesome destructive power of the atomic bombs dropped on Hiroshima and Nagasaki in 1945 left many people fearful of the atom and its potential. Deadly accidents at nuclear power plants such as occurred in Chernobyl, Russia, and Fukushima, Japan, have also sowed public doubts about nuclear energy. As science delves ever deeper into the subatomic world, society will probably remain wary about where these discoveries may lead.

CHAPTER ONE

An Ancient Idea Becomes a Modern Theory

Modern science is inconceivable without the atomic theory, yet the idea that the universe is made up of tiny bits of matter is actually quite old. The first statement of this idea comes from the ancient Greek philosopher Democritus, who lived in the fifth century BCE. Democritus developed the theory from ideas that originated with his mentor Leucippus. According to the theory, the universe is made up of tiny particles that move about in an infinite void. Matter, like clumps of earth, can be broken into pieces because of the empty space between particles. Each piece in turn can be broken in two, and then broken again, and so forth. Finally there must come a point when no empty space remains, and the tiny piece can no longer be divided. Democritus describes this ultimate particle of matter as *atomos*, a Greek word meaning *indivisible*. Thus the smallest building block of matter is the atom, of which there are an infinite number in the universe.

Democritus's atomic theory was purely a thought experiment. He had no means to test his ideas in the real world. Nevertheless his conception proved to be remarkably close to the truth. Matter is indeed composed of atoms, countless trillions of them. Democritus's further speculations were wide of the mark. He believed that each substance owed its different properties to the shape of its atoms and how they fit together. For example, water had smooth atoms that flowed easily, iron had hard, tightly packed atoms, and fire had thorny atoms that pricked and burned. Democritus also was mistaken in his belief that atoms were solid and indestructible.

Democritus's ideas about matter had limited influence. The main problem was that the Greek philosopher Aristotle, born about fifty years after Democritus, dismissed his predecessor's atomic hypothesis. Aristotle believed that all matter was made up of some combination of four elements: earth, air, fire, and water. For him, matter was infinitely divisible. He insisted that a void—known today as a vacuum—could not exist. Aristotle was considered the greatest thinker of the ancient world, and in later centuries the Catholic Church endorsed his ideas about matter to the point that they were considered almost equal to scripture. Rival theories were denounced as heresy. The Church's authority in the Middle Ages ensured that Christian society believed in the Aristotelian model.

The Atomic Theory Is Revived

It took almost two thousand years for Democritus's ideas about the atom to be revived. Natural philosophers—the forerunners of today's scientists—began to examine matter in a systematic way in the 1600s. This scientific revolution challenged Aristotle's worldview and led to new hypotheses about matter and different substances. In 1649 Pierre Gassendi, a professor of mathematics in Paris, France, published a book on the Greek philosopher Epicurus that revived the Greeks' ideas about atoms and the void. Gassendi accepted that tiny bodies called atoms make up the universe. He theorized that the shapes of atoms—round, squat, spiked, or elongated—determine their qualities. He believed atoms bond together with a sort of hook-and-eye arrangement to form "molecules"—a word that Gassendi may have coined. He thought that the hardness of a substance depends on the amount of empty space around its atoms. For example, he believed that vapor results from an increase in the distance between atoms in a liquid. He even proposed that light is made of particles that come from bright objects like the sun and stars. Gassendi's ideas about atoms led him to theorize about the role they play in the world's unfolding creation: "While [the atoms] are moving in various ways and meeting, interweaving, intermingling, unrolling, uniting, and being fitted together, molecules or small structures similar to molecules are created, from which the actual seeds are constructed and fashioned."[4]

The ancient Greek philosopher Aristotle (pictured) believed that all matter consisted of some combination of earth, air, fire, and water. His idea, though incorrect, influenced scientific thinking for centuries.

Gassendi's published work held an element of danger. In 1624 the Parliament of Paris had declared that opposing Aristotle's positions was punishable by death. Yet Gassendi's work, which was profoundly anti-Aristotelian, was allowed to proceed, partly because he had influential friends but also because he differed from the Greek atomists on one important point. Where they declared that atoms had always existed, Gassendi said atoms were made by God. This freed his atomic theory from accusations of atheism (a disbelief in the existence of God). Gassendi's focus on atomic theory helped sweep away Aristotle's errors and prepared the ground for crucial experiments to come.

The Vacuum Pump and the Study of Gases

One of Gassendi's main disagreements with Aristotle was about the void, or empty space between atoms. Aristotle—and the Catholic Church that supported his views—denied that a void could exist. However, experiments in Gassendi's lifetime offered proof of the void's existence. In 1643 the Italian scientist Evangelista Torricelli invented the barometer, an instrument used to measure air pressure and predict changes in the weather. Torricelli's invention featured an empty gap above a column of mercury in a sealed tube. The gap was a vacuum. These and later experiments also provided clues about atoms and the nature of matter.

The movement toward careful experimentation was essential for proving aspects of the atomic theory. The scientific method, as it came to be known, is a procedure for creating hypotheses and testing them with experiments that establish facts and can be reproduced by others. Muslim scholars between the tenth and fourteenth centuries were the first to use this procedure, and it was vital to Western science from the Renaissance forward. As the great English scientist Isaac Newton wrote to a French correspondent in 1672: "For the best and safest method of philosophizing seems to be, first to inquire diligently into the properties of things, and establishing those properties by experiments and then to proceed more slowly to hypotheses for the explanation of them. For hypotheses should be [employed] only in explaining the properties of things, but not assumed in determining them; unless so far as they may furnish experiments."[5]

One of the first great experimenters on vacuum—a space devoid of atoms—and different states of matter was the Irish chemist Robert Boyle, whose best work was done in the latter 1600s. Using a vacuum tube invented by his assistant, Robert Hooke, Boyle investigated a vacuum and its effect on pressure, sound, and motion. For example, Boyle dropped a feather and a piece of lead inside a vacuum tube and watched them hit bottom at the same instant. This proved the Italian scientist Galileo's theory that all objects in a vacuum fall at the same rate. Boyle extinguished a candle flame by pumping air out of the vacuum tube, establishing that combustion requires air. He rang a bell inside the tube then silenced it by pumping out the air, proving that sound does not travel in a vacuum.

Boyle also performed important experiments with gases. He found that the volume of a gas in a confined space varies inversely with the pressure of the gas, a discovery known as Boyle's law. This is the basic idea behind a syringe, and it is also useful to scuba divers who must compensate safely for changes in air and water pressure. Boyle's law essentially proves that gases are made of free-flowing particles with empty space around them. This enables gases to compress or expand as the space decreases or increases. Boyle describes these tiny particles as corpuscles in his own writings, but it is clear that he is referring to atoms and molecules. Boyle hypothesized that everything in nature was the result of particles in motion colliding. Like Gassendi, he was at pains to reject Aristotle's notion of all matter being a combination of only four elements. In its place Boyle offered a new definition of an element. "I now mean by elements," he writes in 1680, "certain primitive and simple, or perfectly unmingled bodies; which not being made of any other bodies, or of one another, are the ingredients of which all those perfectly mixed bodies are immediately compounded, and into which they are ultimately resolved."[6] Boyle perceived that an element was something that could not be broken down by chemical means, that was not a compound or mixture of other substances. This insight would lead eventually to the periodic table of elements and the concept of atomic structure. It was the starting point of modern chemistry.

From Phlogiston to Oxygen

Boyle's work led other scientists to experiment with gases and chemical reactions. One phenomenon that puzzled scientists was combustion. When a piece of paper burns, the remaining ashes have much less mass than the original paper. Some thought the burning material released an element that they called phlogiston. An English

Alchemy and Chemistry

Alchemy was a mystical pursuit dating to the Middle Ages and even to antiquity. Its practitioners combined chemistry, magic, and mysticism in an attempt to transmute, or change, one element into another. Alchemists sought the philosopher's stone, a substance that was supposed to change base metals into gold. Literature and popular culture tend to view alchemy as the province of cranks and charlatans. The title character of Ben Jonson's 1610 play *The Alchemist* is nothing but a con man. Yet alchemy played a surprisingly important role in the origin of modern chemistry and the atomic theory.

The scientific giants Isaac Newton and Robert Boyle both investigated alchemy and believed in its potential. Throughout his career Boyle sought the philosopher's stone, and one of his first books describes what he thought was a method for changing gold into silver. Boyle's experiments in reducing substances to their constituent elements relied in part on alchemists' examples. This work caused him to defend the idea of atomism. Boyle wrote his notes related to alchemy in code—not because he feared ridicule but rather because he dreaded such powerful knowledge falling into the wrong hands. His published work also betrays the influence of alchemy. As the science historian John Christie remarks, "[Boyle] says there are some processes he cannot tell the reader about. He was keeping certain alchemical processes secret. All of this leads me to say that Robert Boyle was an alchemist."

Quoted in *Times Higher Education*, "Boyle the Alchemist," September 27, 1999. www.timeshigher education.co.uk.

clergyman named Joseph Priestley heated mercury oxide to release mercury vapor and what he thought was phlogiston. Priestley noticed that breathing in the colorless gas made him feel light-headed. He also found the gas could be used to ignite a smoldering candlewick. What Priestley thought was "dephlogisticated air" was actually pure oxygen.

The French chemist Antoine Lavoisier repeated Priestley's experiments by heating various metals to form crusts of metal oxide. However, Lavoisier was more meticulous. He employed closed tubes and flasks and carefully collected and weighed the gases and crusts of oxide produced. He found that the oxide crust weighed more than the original metal had before heating. Rather than releasing phlogiston, burning a metal *added* weight from somewhere. Lavoisier discovered that some portion of air was drawn into a substance as it burned. He also found that not all of the air in the flask was involved in the combustion. He finally realized that air must be made up of a mixture of gases, with oxygen the part involved in combustion. The other main gas in the mixture would come to be called nitrogen. While the idea of phlogiston persisted for a while, its existence had basically been disproved. Lavoisier (along with some of his contemporaries) made another breakthrough when he discovered that water is not an element but a compound of hydrogen and oxygen. In 1789 Lavoisier attempted to list the basic elements—and ended up including many that later proved to be compounds. Nevertheless, these hints about how atoms bond led other scientists to make further investigations of atoms and elements.

John Dalton's Atomic Theory

While much important work was being done on isolating elements and compounds, a more complete theory about atoms appeared from an unlikely source. John Dalton was an English chemist and teacher who came from a humble Quaker background. Dalton's first insights about atoms resulted from his work on meteorology and air pressure.

By combining gases in a container, he discovered that the total pressure exerted by the mixture is the sum of all the individual pressures. This suggested to him that each gas has its own weight and pressure. In experiments with the solubility of gases in water, he found that hydrogen, the lightest gas, does not dissolve easily and that heavier gases, such as carbon dioxide, are much more soluble. Dalton theorized that this difference was due to the different weights of the atoms. If atoms have different weights, he reasoned, then they cannot be identical particles of matter, the same for every element. Indeed each element must consist of its own unique atoms.

English chemist John Dalton (pictured) reasoned that every element consists of its own unique atoms. He then calculated and listed the atomic weights of all twenty-one elements known in his time.

Pursuing this idea, Dalton set about calculating the relative weights of different atoms. This was a large challenge, since Dalton could not see the atoms he wanted to measure. Dalton realized the importance of this effort, as he writes in his 1808 book *A New System of Chemical Philosophy*: "Now it is one great object of this work, to show the importance and advantage of ascertaining the relative weights of the ultimate particles, both of simple and compound bodies, the number of simple elementary particles which constitute one compound particle, and the number of less compound particles which enter into the formation of one more compound particle."[7]

Measuring an Atom

As atomic theory revived in the 1600s, thinkers began to speculate about the number and size of atoms. Certainly they were invisibly tiny, but there had to be a way to describe their size with more precision. One of the first attempts to deal with this idea came in a 1646 book by a French physician named Johann Magnenus. He reasoned that, upon burning incense in a cathedral, one atom of incense reached his nostril in the first instant he could smell it. He then did a number of calculations based on the diffusion of incense in the sanctuary. Magnenus's math proved meaningless, but he did make clever speculations about the relative sizes of atoms.

A more substantial approach took shape in 1811, when Italian chemist Amedeo Avogadro published his ideas about the equal number of particles in gases of equal volume and pressure. This led to speculations about measuring the number of molecules in a given weight of a substance, called a mole. The measurement became known as Avogadro's number. Eventually chemists linked the number to the mass of one atom of carbon 12. Avogadro's number ultimately was calculated to be 6.022×10^{23}. This number of carbon 12 atoms weighs exactly 12 grams. Thus Avogadro's number is probably greater than the number of stars in the universe (estimated to be 10^{22}). This indicates how tiny atoms and molecules truly are.

Thus Dalton attempted to find the weights of atoms (ultimate particles) and molecules (compound particles). He could not place atoms on a scale, but he could use the results of other scientists' work as tools. The French chemist Joseph-Louis Proust had discovered the law of definite proportions, which states that the atoms of a compound always combine in the same proportion by weight. For example, carbon dioxide always consists of three units of carbon and eight units of oxygen. Dalton started with the lightest known element, hydrogen, and assigned it a weight of 1. Then, using Proust's law, he measured the amounts of other elements in compounds with hydrogen and assigned them whole-number weights according to their ratios in the compounds. In the course of his work he affirmed that atoms combine in whole-number ratios; there is never half an atom involved. Overall,

Dalton used this method to calculate and list the atomic weights of all twenty-one elements known in his time.

Having marshaled proof about atoms and their properties, Dalton was ready to expand his ideas into a complete theory. In his 1808 book he sets out the main points of his atomic theory, which in essence say:

- All matter consists of small, indivisible particles called atoms.
- Atoms cannot be created or destroyed.
- All atoms of the same element have identical weights, but atoms of different elements have different weights.
- In chemical reactions, atoms combine in simple whole-number ratios.
- Atoms do not change when they become part of chemical compounds.

Despite some errors and oversimplifications, Dalton's points are close to the modern understanding of atoms. His chief accomplishment was to codify what was known about atoms and give it a mathematical basis. However, his atomic theory did not immediately set the scientific world on fire, let alone society in general. As science historian John Gribbin observes, "Many people found it hard, sometimes on philosophical grounds, to accept the idea of atoms (with the implication that there was nothing at all in the spaces between atoms) and even many of those who used the idea regarded it as no more than . . . a tool to use in working out how elements behave as *if* they were composed of tiny particles, without it necessarily being the case that they *are* composed of tiny particles."[8] Dalton's remarkable work did not go unrecognized. He won many honors during the remainder of his life, and more than forty thousand people marched in his funeral procession in Manchester, England. Yet it would be a full century after his landmark publication before actual proof of the atom's existence emerged.

Atoms, Molecules, and the Periodic Table

Soon after Dalton published his atomic theory, other scientists worked to develop it further. The French chemist Joseph Gay-Lussac made a discovery that seemed to affirm Dalton's theory. Gay-Lussac's law of combining volumes stated that gases react with each other in whole number ratios. The Italian chemist Amedeo Avogadro saw this law's potential to explain molecular combinations. In 1811 Avogadro proposed that equal volumes of gases at identical conditions of temperature and pressure contain equal numbers of particles—that is, molecules. Avogadro reasoned further that simple gases are not made of single atoms but instead are compound molecules of two atoms or more. This led him to see that the molecular makeup of water is not HO as Dalton believed, but H_2O. In other words, a water molecule consists of two atoms of hydrogen and one atom of oxygen. For various reasons, Avogadro's diatomic (two-atom) hypothesis lay dormant for a half century after he published it.

The periodic table of elements, familiar to chemistry students worldwide, was devised by Russian scientist Dimitri Mendeleev. Mendeleev arranged the elements in rows and columns according to atomic weights and various properties.

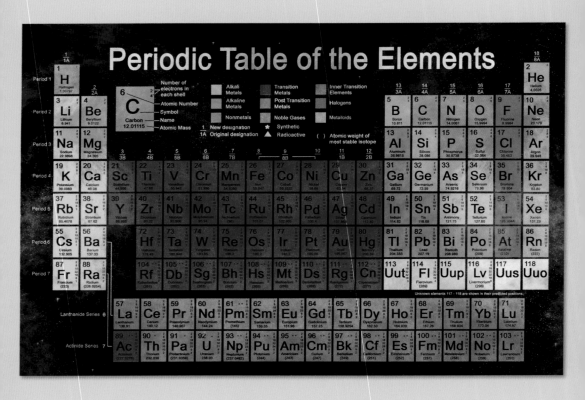

In the ensuing decades, scientists continued to isolate elements and identify their atomic weights and chemical properties. In 1869 Dmitri Mendeleev, a professor at the University of St. Petersburg in Russia, compiled all the known information about the elements in order to create a table of elements for a chemistry textbook. Mendeleev arranged the elements in rows and columns, with hydrogen, the lightest element, in the upper left corner. In each row elements were arranged by increasing atomic weight. Each column contained elements with similar properties, such as valence, or the ability to combine with other elements. Mendeleev noticed that as atomic weight increased, similar chemical properties arose at intervals, or periods. Thus he named his chart the periodic table. A few years later chemists discovered previously unknown elements whose existence and properties were predicted by open slots in Mendeleev's table. Chemistry itself was becoming a precise discipline that created undreamed of new products and industries. The stage was set for astounding discoveries about atoms and molecules in the next century.

CHAPTER TWO

The Atomic Theory Develops

As the nineteenth century waned, new revisions of John Dalton's atomic theory appeared. Dalton's inability to probe subatomic structure left blank spots in his concept of the atom and provided opportunities for later scientists. Experiments with electromagnetic fields, cathode rays, and improved vacuum tubes called Crookes tubes yielded surprising results and fresh discoveries. Scientists were beginning to plumb the mysteries of the subatomic world.

The Atom and Brownian Motion

While some scientists proceeded with work that expanded Dalton's theory, his basic proposal about the existence of atoms had yet to be affirmed. For most reputable scientists, Dalton's theory—that the universe consists of an almost infinite number of tiny particles held together by some strange force to form all matter—was hard to accept. It was clever and elegant as far as it went, but some sort of solid proof was required. In 1827 experimental proof did indeed appear, and in the unlikeliest of circumstances. The first hard evidence for the atom's existence came from an expert on plants.

In 1827 the Scottish botanist Robert Brown placed fine grains of pollen—each less than half a hundredth of a millimeter in diameter—into water and viewed them through a microscope. Brown observed that the pollen grains seemed to jump and skitter about in an odd fashion, making random movements with no apparent impetus. It was soon discovered that the movement of the pollen grains was not due to their being alive but rather was caused by the action of molecules in the water pushing the grains about. While Brown himself did not correctly explain the movement of the pollen grains, it is named in his honor: Brownian motion.

The full implication of Brown's discovery had to wait nearly eighty years, a period during which many scientists remained skeptical of atomic theory as a practical matter. In 1905 a German-born patent clerk and scientist named Albert Einstein published four groundbreaking papers, one of which concerned Brownian motion and the measurement of molecules. Einstein had studied the phenomenon of Brownian motion more closely than anyone before him. He produced a formula, based on the statistical averages of the pollen grains' random movements, to calculate the number of water molecules per square inch—a calculation that proved amazingly accurate. Einstein also offered proof that it was the motion of molecules bumping the pollen grains from all sides in a random fashion that caused their movement. A French scientist, Jean Perrin, carried out the tedious work of measuring this random movement and verifying Einstein's ideas. The scientific community acknowledged the importance of Brownian motion in affirming the existence of atoms and molecules. Perrin eventually won the Nobel Prize for his efforts. "I would have considered it impossible to investigate Brownian motion so precisely," wrote a grateful Einstein to Perrin. "It is a stroke of luck for this subject that you have taken it up."[9] As for Einstein's formula, which was dubbed the random walk, it would prove useful in other investigations of the atom's properties.

The Plum-Pudding Atomic Model

One individual who was intrigued by atoms and electromagnetism was the English physicist Joseph John Thomson. In 1897 he conducted a series of experiments with cathode ray tubes. Cathode rays were known only as mysterious emissions of fluorescent light that appeared when voltage was applied to two metal electrodes in a glass vacuum tube. The rays came from the cathode, or negatively charged wire. Exhibitions of this colorful phenomenon—flowing patterns of light in a glass tube—delighted crowds in the late 1800s. The cathode ray tube would soon lead to another crowd-pleasing invention: television.

Thomson found that cathode rays not only have a negative charge, they also bend as they pass through an electromagnetic field. He realized that cathode rays consist of particles with a negative charge. By calculating the charge-to-mass ratio of the rays, Thomson made a remarkable discovery. Each particle in cathode rays seemed to be smaller than a hydrogen atom. This refuted Dalton's idea that an atom is the smallest indivisible piece of matter. It also suggested that an atom is not a solid, featureless ball as Dalton and his contemporaries assumed but instead has its own particles. Thomson called the particle he discovered an *electron*. He proposed that negatively charged electrons are scattered throughout the positively charged sphere of an atom like blueberries in a muffin. Indeed his idea was called the plum-pudding atomic model, for the fruit-filled dessert favored by the British.

WORDS IN CONTEXT

cathode rays
Streams of electrons observed in a vacuum tube.

Ernest Rutherford and the Planetary Atomic Model

While Thomson's discovery of the electron soon gained acceptance, other scientists set out to improve upon his atomic model. The most successful of these by far was one of Thomson's best students, a rural New Zealander named Ernest Rutherford. Unlike Thomson, his mentor, who was known for his clumsiness in the laboratory, young Rutherford was adept at performing experiments. To modern scientists Rutherford's own methods might seem haphazard and crude, but they proved very effective.

Rutherford and Thomson were intrigued by recent discoveries of certain special kinds of radiation. In 1895 German physicist Wilhelm Roentgen had found that sending electric current through a Crookes tube created powerful rays capable of illuminating a nearby screen covered in fluorescent material even though the tube was sheathed in black paper. Roentgen named these rays X-rays. A year later French physicist Henri Becquerel sought to produce X-rays on his own. Thinking the rays were connected to phosphorescence, Becquerel tried to create them from phosphorescent salts. One such salt, uranium, could fog a sheathed photographic plate in a dark room. Becquerel had discovered radioactivity. The name came from another physicist in France, Marie Curie, who likened uranium's emission of energy to a radio transmitter.

Rutherford set about investigating radioactivity. He found that the radiation produced by radioactive elements comes in two forms, which he named *alpha* and *beta*. A third kind of radiation was soon discovered and named *gamma*. Rutherford also made a discovery that

The most famous experiment performed by New Zealand scientist Ernest Rutherford (pictured in his laboratory at McGill University in Canada in 1905) led to a new model of the atom. It became known as the planetary atomic model.

The Atom in Popular Culture

Scientific breakthroughs often take time to seep into popular culture. Edward Rutherford announced his planetary model of the atom in 1911, but it took decades for this model to become widely known to the public. By the 1950s the familiar picture of the atom with its symmetrical orbiting electrons was the ultimate symbol of modern life. Almost any business or product could seem up-to-date by including the word *atomic* in its name or employing the atom symbol in its logo. The period was often referred to as the Atomic Age.

The symbol appeared on everything from furnishings and appliances to toys and knickknacks. RCA sold vacuum tubes for radios and televisions packaged with the atom symbol. Designers used the atom motif for wallpaper and rugs. A popular chemistry set for children was called the Atomic Energy Lab. A label on the box said it contained Safe Radioactive Materials. Belgian engineers built a 330-foot atom-like structure out of stainless steel for the 1958 World's Fair.

The atomic fad also spread to magazines and movies. A Japanese cartoon character was called the Mighty Atom. DC Comics introduced its own Atom, a superhero who could shrink to miniscule size. Hollywood produced movies such as *The Atomic Monster* and *Creature with the Atom Brain*. Not even anxiety about the atomic bomb could dampen enthusiasm for the atom. A 1951 issue of *Motor Trend* magazine trumpeted "Tomorrow's Atom Car!" on its cover, with a mushroom cloud rising in the background.

Moxie and Suzy, "Atomic Age Design—Beauty and the Bomb," *Retropedia* (blog), January 23, 2013. http://revivalvintagestudio.blogspot.com.

seemed to hearken back to the alchemists. Radioactive decay—the release of alpha and beta particles by atoms—actually changes one kind of element into another.

However, it was Rutherford's most famous experiment that led to a new model of the atom. In the experiment, he sent a beam of positively charged alpha particles into a razor-thin piece of gold foil. An assistant sitting in the darkened room counted the brief flashes of light, or scintillations, caused by particles hitting a detection screen. Rutherford expected the particles to pierce the foil in mostly straight lines, with only a few deflected slightly by the electrons' negative charge. However, his assistant reported more deflection than expected. In fact, a few of the particles did not go through the foil at all but actually bounced back toward the source of the radiation. As Rutherford admitted years later, "It was quite the most incredible event that has ever happened to me in my life. It was almost as incredible as if you fired a 15-inch shell at a piece of tissue paper and it came back and hit you."[10]

Rutherford deduced that while most of the particles passed through the foil as he expected, some collided head-on with the nucleus of one of the few hundred atoms they had to pass through. When this occurred, the positively charged particle was evidently repulsed by the positively charged nucleus. Rutherford realized this meant most of the atom's mass was concentrated in its nucleus. In 1911 Rutherford used his findings to propose a new model of the atom to replace the plum-pudding model. He stated that an atom is composed of a tiny nucleus with a positive charge orbited by even tinier electrons with a negative charge. Rutherford estimated that the radius of the nucleus was ten thousand times smaller than the radius of the entire atom with its orbiting electrons. For example, if the tiny nucleus were enlarged to the size of a marble and placed on the pitcher's mound at Yankee Stadium, the nearest electron would be in the top row of the upper deck. In other words, an atom is made up mostly of empty space, and so is everything that people think of as solid matter. Since Rutherford's conception was similar to the way planets orbit the sun, it became known as the planetary atomic model.

Atoms and Quantum Mechanics

History shows that Rutherford's model was a vast improvement over past conceptions of the atom. Yet just as Dalton's and Thomson's atomic theories were subjected to scrutiny and revision, Rutherford's model was soon found to have its own flaw. The problem had to do with the known laws of physics. Electrons circling around the nucleus should lose energy as they emit radiation, until they slow down and spiral into the nucleus. According to the laws of physics, Rutherford's atoms could not exist.

The solution to this problem lay with a revolution in physics that came to be called quantum mechanics. It was based on the ideas of three physicists, Germany's Max Planck, Albert Einstein, and Denmark's Niels Bohr. In 1900 Planck was a professor at the University of Berlin where he was studying thermodynamics, or the science of heat and energy. He was interested in how objects radiated energy when heated. Growing more puzzled with results that refused to obey the laws of physics, Planck finally concocted a wild theory: "The whole procedure was an act of despair because a theoretical interpretation had to be found at any price, no matter how high that might be. . . . I was ready to sacrifice any of my previous convictions about physics."[11]

He proposed that radiated energy does not flow as a continuous wave but rather is emitted in tiny, discontinuous units or packets. Planck invented the name *quanta* for these tiny chunks of energy; a *quantum* was a single chunk. Planck saw quanta as the basic units of energy, or as he described them, "the pennies of the atomic world."[12] Like units of money, quanta appear only in certain whole-number multiples, never in fractional states. The energy of each quantum is equal to the frequency of radiation multiplied by a constant value, which Planck calculated. This value is known as Planck's constant. Oddly enough, Planck initially did not believe he had uncovered an important truth about energy, but instead that he had found only a mathematical trick to explain certain phenomena.

Planck's proposals about quanta baffled other scientists, who could not reconcile his findings with classical physics. It was left to Einstein to validate Planck's findings. Einstein discovered that electromagnetic radiation had to consist of quanta—or photons, as the chunks of light energy came to be called. At first Planck himself ridiculed Einstein's conclusions, but he later came to accept them, as did the rest of the scientific world. Einstein's work on quanta earned him the Nobel Prize for physics in 1921. Atomic theory would soon be recast to fit with the quantum world.

Niels Bohr and the Quantum Model

Niels Bohr was the first to see the relationship between Planck's quantum ideas and the behavior of electrons. Bohr was a young physicist

The revolution in physics known as quantum mechanics had its basis in the ideas of three scientists: Max Planck, Niels Bohr, and Albert Einstein (pictured). Through their work, scientists gained a deeper understanding of electrons—and atoms in general.

who had worked with both Thomson and Rutherford in England. Rutherford was not only his mentor but virtually a father figure. Back in his native Copenhagen, Bohr began to study the problems with Rutherford's atomic model. Classical physics insisted Rutherford's version could not exist. As Bohr recalled years later, "It was clear, and that was *the* point about the Rutherford atom, that we had something from which we could not proceed at all in any other way than by radical change."[13] His key insight was that quanta—chunks of energy—could explain how electrons orbit the nucleus of an atom without losing energy and becoming unstable.

The Quirks of Quarks

The last fifty years have seen continual revisions and additions to atomic theory. Many of these discoveries are due to powerful particle accelerators capable of accelerating intense beams of electrons up to incredible energies. By smashing particles together, physicists create other particles, some lasting for only a fraction of a second. The existence of these new particles reveals that neutrons and protons are themselves formed from smaller bits of matter. One of these fundamental particles is the quark.

Two physicists, American Murray Gell-Mann and Russian-born George Zweig, proposed the existence of quarks in 1964. Gell-Mann chose the name from a nonsense word in James Joyce's novel *Finnegans Wake*. For years other scientists refused to believe that quarks really existed. However, experiments using particle accelerators eventually revealed all the types of quarks that Gell-Mann had predicted. There are six kinds of quarks, generally considered in pairs named up/down quarks, charm/strange quarks, and top/bottom quarks. The top quark was the last to be discovered, in 1995. Proof of quarks' existence is based solely on mathematical analysis of complex experiments. As Gell-Mann explains, "You can't see [quarks] directly. They have some unusual properties, and that's why it was difficult for people to believe in them at the beginning. And lots of people didn't. Lots of people thought I was crazy. Quarks are permanently trapped inside other particles like neutrons and protons. You can't bring them out individually to study them. So they're a little peculiar in that respect."

Quoted in Susan Kruglinski, "The Man Who Found Quarks and Made Sense of the Universe," *Discover*, March 17, 2009. http://discovermagazine.com.

Bohr theorized that what Rutherford had described as the electrons' orbits are actually energy levels. These levels determine the fixed paths in which electrons can move around the nucleus. An electron does not emit energy as long as it remains on one path or level. When an electron moves from a lower energy level to a higher one, it absorbs a quantum of energy. This is called an excited state. When it moves back to a lower energy level, it gives off a quantum of energy in the form of a photon. The energy levels can be thought of as shells, like the layers of an onion. These layers correspond to Planck's quanta. An electron must always be on one layer or another. It cannot be in between layers.

Bohr tested his model by using it to predict the spectra, or different values, of light emitted by different kinds of atoms. Atoms release energy as a photon of light, the color of which depends on the amount of energy released. Bohr found that the different energy levels produced exactly the expected lines and gaps of emission and absorption. As John Gribbin notes, "Quantum physics had explained why, and how, each element produces its own unique spectral fingerprint. The model might be a crazy patchwork of ideas old and new, but it worked."[14] Now, however, things were about to get really strange.

Heisenberg's Uncertainty Principle

Despite its ability to predict light spectra correctly, the Rutherford-Bohr atomic model confounded many scientists. Some dismissed the whole idea as nonsense. Yet many others joined in the search for more answers. Bohr himself used quantum theory to explain why elements differ in their abilities to form compounds, or elements that are chemically joined. In the meantime, the French scientist Louis de Broglie introduced a new complexity to atomic theory. He suggested that matter, in the form of electrons, behaves sometimes like particles and sometimes like waves. In 1923 the young Austrian physicist Erwin Schrödinger made use of de Broglie's insight. Schrödinger theorized that an electron did not orbit around the nucleus but rather oscillated around it like the waves in a jump rope. This would explain why only certain wavelengths fit into an orbit. Schrödinger came up with a mathematical equation based on this wave idea that explained all the facts about Bohr's atomic model. A German physicist named Werner Heisenberg employed a different mathematical approach along with the idea of electrons as particles to do the same thing. Their efforts placed quantum mechanics on a solid theoretical basis. They also showed that matter is apparently wavelike and particle-like at the same time.

In 1927 Heisenberg noted a further oddity about electrons. He claimed that it is impossible to know simultaneously the location of an electron and its velocity. Merely shining a weak beam of light on an electron causes it to change speed, so no observation works. An observer can know either the location or the speed with some precision, but never both at the same instant. The best one can do is predict where an electron is likely to be. Heisenberg's idea, called the uncertainty principle, changed the atomic model once more. Electrons are now thought of as oscillating in cloudlike areas at different distances from the nucleus. At any given instant an electron might be anywhere on its wavelike path. With this uncertainty in mind, Heisenberg gave up on depicting the atom as a visual model.

Some scientists rebelled against the findings of quantum mechanics. One of these was Einstein himself. While he had enormous respect for the younger physicists and their accomplishments, he could not accept that the universe is built on uncertainty. Einstein subscribed to Newtonian ideas of physics, in which all natural forces can be explained and predicted. "God does not play dice with the universe,"[15] Einstein declared many times. To the end of his life he searched for an overriding theory that would resolve the randomness of the quantum atomic model. He never found it.

Atomic Theory Today

The uncertainty principle and randomness may be essential to atomic theory, but most nonscientists today have a much simpler idea about atoms. Generally this is some version of the Rutherford-Bohr atom, with a nucleus formed of neutrons and protons orbited by electrons. This atomic model can provide a misleading picture of the universe as an orderly, explicable system. Bohr, Schrödinger, and Heisenberg knew differently almost a century ago. And in the succeeding years physicists have delved ever deeper into atomic structure—splitting the nucleus, locating increasingly smaller particles, discovering virtual particles that exist for only a fraction of a second.

Despite its crucial role in modern technology, quantum theory remains mostly a mystery to the general public. Moviegoers have become acquainted with its notions of time travel and multiple realities, but the bizarre nature of today's atomic theory is mostly ignored. People rely every day on smart phones, lasers, MRI machines, and

The Structure of the Atom

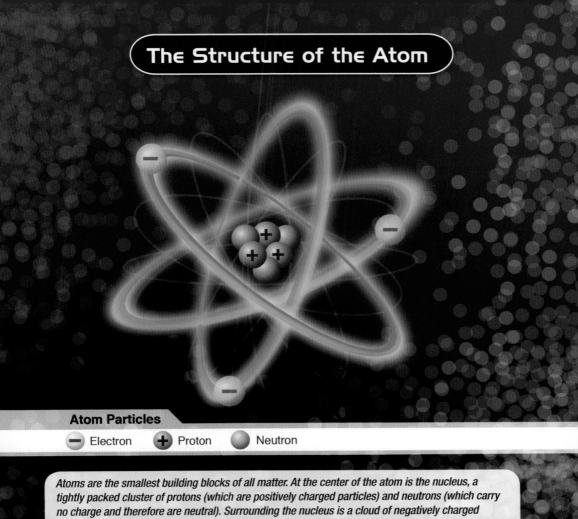

Atom Particles

- Electron
- Proton
- Neutron

Atoms are the smallest building blocks of all matter. At the center of the atom is the nucleus, a tightly packed cluster of protons (which are positively charged particles) and neutrons (which carry no charge and therefore are neutral). Surrounding the nucleus is a cloud of negatively charged electrons. Within each atom, the number of protons and electrons is the same, resulting in a neutral charge. Scientists know which element an atom is by its atomic number, which is the number of protons in its nucleus. For instance, a hydrogen atom has one proton in its nucleus while an oxygen atom has eight protons in its nucleus. This atomic number is how elements are organized in the periodic table of elements.

Source: Atomicarchive.com, "Nuclear Fission: Basics," 2013. www.atomicarchive.com.

many other inventions that make use of the quantum theory of the atom. Some might say the indeterminate nature of the latest atomic theory has contributed somehow to feelings of confusion and helplessness in today's world. But perhaps that question is best left to philosophers and social scientists.

CHAPTER THREE

Nuclear Energy and the Atomic Age

In July 2014 authorities in Japan took steps to reopen the Sendai nuclear power plant in the southern part of the country. The plant was to be the first Japanese nuclear facility to restart after the Fukushima disaster of 2011. At Fukushima, a tsunami caused by a massive earthquake led to the meltdown of three out of six nuclear reactors and the release of dangerous amounts of radioactive material into the air and water. After the accident the Japanese government shut down all forty-eight nuclear reactors in the country. Residents near the sites have expressed fears that new safety regulations are still not adequate. Many have signed petitions and led protests hoping to keep the plants closed. Yet Prime Minister Shinzo Abe believes nuclear energy is cheaper than fossil fuels and therefore essential to Japan's economy. Naoto Kan, Abe's predecessor as prime minister, disagrees. "This government has not learned the lessons of Fukushima," says Kan. "Japan was on the brink, but now we want to go back to nuclear for economic reasons. But what happens to the economy if another disaster hits?"[16]

The controversy is typical of the strong feelings, both pro and con, raised by nuclear power. Its potential as a source of relatively clean energy is limitless, yet fears of meltdown and nuclear radiation make it a hard sell in many places. The ability to split the atom—and the resultant power plants, radioactive waste, and nuclear missiles—can seem as much a curse as a blessing in the modern age.

Splitting the Nucleus

Blessing or curse, the idea of nuclear energy began as a series of further discoveries about the structure of an atom. In 1919 Ernest Rutherford discovered that the hydrogen nucleus is a basic component of every atom. Rutherford called the hydrogen nucleus a proton. Each proton in an atom has a positive electrical charge, equal but opposite to the electron's negative charge. The number of protons in a nucleus also determines what chemical element the atom is and its place in the periodic table. This number is referred to as the atomic number.

Rutherford's discovery raised the question of how tightly packed, positively charged protons keep from repelling each other and flying apart. In 1932 the English physicist James Chadwick solved the problem when he suggested yet another particle in the nucleus: the neutron. This particle had a neutral charge and served to buffer the electrical forces between protons, thus holding the nucleus together. With his typical modesty, Chadwick announced his findings in a paper titled "Possible Existence of Neutron." Werner Heisenberg soon affirmed Chadwick's discovery. However, it was other physicists who saw startling possibilities for the neutron.

With its neutral charge, the neutron was perfect for bombarding the nuclei of atoms. It could smash into a nucleus like a bullet, with no drag from the repelling force of similarly charged particles. For the next several years after Chadwick's discovery, teams of physicists in several countries set about firing neutrons at every known element. They reported the creation of hundreds of new radioactive isotopes. In 1934 a team led by the Italian physicist Enrico Fermi published the results of their bombardment of uranium. They could not explain the different radioactive substances they produced. Something unexpected was occurring, an effect that became known as the Uranium Problem.

In 1938 a trio of German physicists solved this problem. They discovered that bombarding uranium with neutrons produced isotopes of the lighter elements barium and krypton. The neutrons had actually split the uranium nuclei in two. The mass that was lost had been converted into a tremendous amount of energy. The trio—Otto Hahn, Lise Meitner, and Fritz Strassman—had changed the world with their discovery of nuclear fission: the splitting of the atom.

Computer artwork depicts the splitting of the atom, or nuclear fission. Scientists Otto Hahn, Lise Meitner, and Fritz Strassman discovered that an enormous amount of energy is created when the atom's nucleus (center) splits into smaller fragments with less mass.

A Chain Reaction

One person who immediately realized what this breakthrough meant was the Hungarian physicist Leo Szilard (pronounced SIL-ahrd). Chadwick's discovery of the neutron had led Szilard to make a number of speculations. He thought that a densely packed mass of fissionable material could create critical mass—a self-sustaining chain reaction. As science writer Paul Parsons describes it: "Each time a uranium nucleus splits in half, it emits more neutrons. And each of these neutrons can then go on to split another uranium nucleus, repeating the process and setting up a self-perpetuating 'nuclear chain reaction.'"[17] Szilard understood that such a chain reaction could result in a controlled release of atomic energy. He also realized that this chain reaction could be the basis for making an atomic bomb. Part of Szilard's inspiration was a 1914 science fiction novel by British author H.G. Wells in which scientists manufacture atomic bombs in an effort to unite the world into one society.

Szilard was still not certain a nuclear chain reaction was possible, but he was alarmed by its destructive potential. On July 4, 1934, he filed for and was granted a patent for his idea of a neutron-based chain reaction. His motive was not profit but caution; he feared the idea falling into the wrong hands. In fact, he tried to hand the idea over to the British government so it could be shielded by intelligence laws. In January 1939, when Szilard learned about the discovery made by Otto Hahn's group, he knew that his idea about a nuclear chain reaction was essentially correct. Prospects for an atomic bomb were more than idle speculation. With the Nazis threatening war in Europe, he decided to act.

The Manhattan Project

Szilard contacted Albert Einstein, whose reputation as the world's greatest scientist lent his words and opinions extra weight. Szilard expressed fears that Hitler's ministers of war were developing an atomic bomb capable of killing untold numbers of innocent people. Einstein himself was a noted pacifist, and he dreaded such a weapon in the hands of any nation, even one of the western democracies. Yet finally, at Szilard's urging and with his help, Einstein wrote a letter to the American president, Franklin D. Roosevelt, warning him about the potential of an atomic bomb. The letter, dated August 2, 1939, reads in part: "This new phenomenon [of a nuclear chain reaction] would also lead to the construction of bombs, and it is conceivable—though much less certain—that extremely powerful bombs of a new type may thus be constructed. A single bomb of this type, carried by boat and exploded in a port, might very well destroy the whole port together with some of the surrounding territory."[18]

The letter went on to suggest sources of uranium ore in the United States that could be used for nuclear research. It also recommended that the US government speed up funding for such research at American universities. Einstein's doubts as a pacifist left the message indirect, yet in essence he and Szilard were pressing Roosevelt to develop an atomic weapon.

The result of Einstein's letter was the Manhattan Project, a top-secret plan to make an atomic bomb for military use. Ironically, US Army Intelligence forbade the hundreds of scientists enlisted for the effort from consulting with Einstein. Officials decided Einstein's left-wing views were too radical for him to have security clearance. Nevertheless, a team of the world's top physicists, including J. Robert Oppenheimer, Niels Bohr, Enrico Fermi, and Richard Feynman helped the Manhattan Project achieve its goal. On July 16, 1945, a test bomb nicknamed Fat Boy was detonated in the New Mexican desert.

At this stage of World War II, Nazi Germany had surrendered and the Japanese military was close to defeat. Yet bloody fighting from island to island in Japan convinced Allied leaders to make a fateful decision. On August 6, 1945, an American B-29 bomber dropped a nine-thousand-pound atomic bomb on the Japanese city of Hiroshima. Three days later another bomb was dropped on the port of Nagasaki. The death toll from the two explosions was staggering. A total of more than one hundred thousand people died in the initial blasts. Many were vaporized instantly, while thousands more perished in fires and collapsing buildings. At least eighty thousand people died in the ensuing years from various related causes, including radiation sickness. World reaction was mixed, with many pleased that the war was finally over and others shocked that the Americans had unleashed such a frightening weapon. Journalists began to refer to the Atomic Age, and the bomb's mushroom cloud became a symbol of modern times.

Nuclear Arms Race

American dominance of the Atomic Age did not last long. In 1949 the Soviet Union tested its own atomic bomb at a remote site in

In September 1945, one month after the United States dropped a nine-thousand-pound atomic bomb on the Japanese city of Hiroshima, a western news reporter surveys a landscape of utter ruin. The Atomic Age had begun.

Kazakhstan. US authorities arrested Klaus Fuchs, a German-born physicist who had worked on the Manhattan Project, for passing secrets about bomb making to the Soviets. In response to the Soviet test, the US rushed to acquire an even more powerful nuclear weapon.

Hungarian-born physicist Edward Teller led efforts to develop a new superbomb, despite opposition from much of the scientific community. "If the Russians demonstrate a super [bomb] before we possess one, our situation will be hopeless,"[19] Teller declared. The new weapon was based on fusion, the opposite of the fission process used in the first atomic bombs. Instead of splitting a nucleus, fusion combines the nuclei of two atoms to form a single heavier atom. The fusion reaction requires extremely high temperatures, which is why the weapon based on it is called a thermonuclear bomb. The more common name is the hydrogen bomb. On November 1, 1952, the United States successfully tested the first hydrogen bomb in the Pacific Marshall Islands. The explosion was seven hundred times more powerful than the atomic bomb and produced a crater more than a mile wide. Three years later the Soviet Union detonated its own hydrogen bomb. In the Cold War between the United States and the Soviet Union, both sides developed huge arsenals of nuclear weapons with missiles to deliver them from long range. Each nation was deterred from striking first by the idea of mutually assured destruction (MAD). If one of the foes attacked with nuclear missiles, the other would respond in kind and both sides were certain to be destroyed.

The American people saw the self-confidence that came from winning World War II change to anxiety at news of the Soviet Union's development of nuclear weapons. During the 1950s ordinary families built bomb shelters in their backyards, and schoolchildren practiced duck-and-cover safety maneuvers at their desks to prepare for a possible missile attack. Tensions reached a fever pitch in 1962 when American spy planes photographed Soviet workers building nuclear missile sites on the island of Cuba, about ninety miles from the coast of Florida. President John F. Kennedy ordered a naval blockade—a ring of ships to stop all incoming traffic. Soviet leader Nikita Khrushchev finally agreed to dismantle the sites in exchange for Kennedy's promise not to invade Cuba.

What some saw as the madness of the nuclear arms race was reflected in movies such as *On the Beach* (1959) and *Dr. Strangelove* (1964). The latter film is a wild satire with the subtitle "How I Learned to Stop Worrying and Love the Bomb." In it Teller is lampooned as a nuclear physicist who smilingly explains the simplicity of a Doomsday Bomb. The film's final image is an American pilot riding a nuclear bomb like a bucking bronco as it falls toward Earth.

Nuclear Power as an Energy Source

While nuclear fission is the basis for the world's most fearsome weapons, it also has peaceful applications. Fission is the idea behind nuclear power plants capable of generating enormous amounts of electricity. In a nuclear reactor, a continuous fission reaction heats water into pressurized steam, which drives a steam turbine. The turbine then generates electricity. A byproduct of nuclear energy is spent uranium fuel or radioactive waste. Managing this waste, which is thermally hot

John Hersey's "Hiroshima"

Immediately after the Allies dropped two atomic bombs on Japanese cities in August 1945, Americans understood very little about what had actually occurred. Most knew only that the explosions had brought an end to the war. This changed a year later with the publication of John Hersey's "Hiroshima." Few pieces of journalism in the twentieth century have done more to influence public opinion. The article by the veteran war correspondent appeared in the *New Yorker* magazine on August 31, 1946. The magazine's editors originally scheduled the article to run as a three-part series but decided instead to devote an entire issue to it.

"Hiroshima" traces the lives of six people who survived the Hiroshima atomic bomb. In precise, unemotional prose, Hersey describes the harrowing aftermath of the explosion. Here he focuses on the efforts of a young surgeon to provide help: "By nightfall, ten thousand victims of the explosion had invaded the Red Cross Hospital, and Dr. Sasaki, worn out, was moving aimlessly and dully up and down the stinking corridors with wads of bandage and bottles of mercurochrome, still wearing the glasses he had taken from the wounded nurse, binding up the worst cuts as he came to them. Other doctors were putting compresses of saline solution on the worst burns. That was all they could do."

Hersey's article brought home in shattering detail the human cost of dropping the atomic bomb. It caused many readers to rethink their attitudes about modern science and the unleashed power of the atom.

John Hersey, "Hiroshima," *New Yorker,* August 31, 1946. www.newyorker.com.

as well as extremely radioactive, is one of the challenges of operating a nuclear power plant.

After World War II many scientists recognized fission's promise as a source of energy. In 1951 a small experimental nuclear reactor built in Arco, Idaho, became the first to generate electricity, albeit a tiny amount. Two years later President Dwight D. Eisenhower announced a program called Atoms for Peace, which outlined plans for developing nuclear energy across the United States. In 1954 the Soviet Union began operating the first full-scale nuclear reactor—called *Atom Mirny* (peaceful atom)—in the city of Obninsk, southwest of Moscow. Soon thereafter Calder Hall in Cumbria, Great Britain, became the world's first commercial nuclear power plant. When the plant closed in 2003 for safety reasons, its staff was still touting its usefulness as a source of clean energy. "This station was the pioneer," says manager John Vlietstra. "Its long and reliable lifetime has proved that nuclear power makes a real contribution to the country's needs for electricity. Since 1956 it has been producing greenhouse-gas–free electricity for around 200,000 homes."[20]

Three Mile Island and Chernobyl

Nuclear power plants like Calder Hall once seemed like the logical answer to the world's energy needs. A pound of highly enriched uranium can generate energy equivalent to a million gallons of gasoline. Yet the hopes for nuclear power suffered a major blow on March 28, 1979. The Three Mile Island reactor located near Middletown, Pennsylvania, underwent a severe core meltdown. The incident began with a mechanical or electrical failure that caused the plant's turbine-generator to shut down. A pressure relief valve then became stuck open, allowing coolant to pour out. According to the report by the US Nuclear Regulatory Commission, "As alarms rang and warning lights flashed, the operators did not realize that the plant was experiencing a loss-of-coolant accident. They took a series of actions that made conditions worse. . . . It was later found that about half of the core melted during the early stages of the accident."[21]

WORDS IN CONTEXT

meltdown
Accidental melting of a nuclear reactor's core.

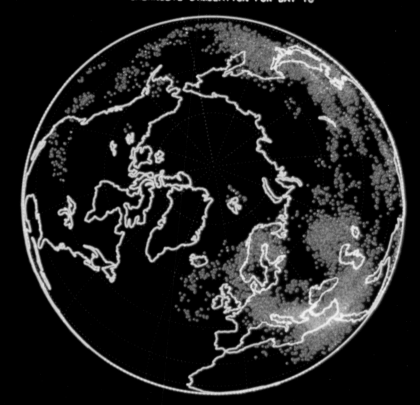

Ten days after the meltdown at the Chernobyl nuclear reactor in Ukraine, the US National Atmospheric Release Advisory Center plots the spread of radioactive material around the northern hemisphere. The disaster was attributed to flawed design and inexperienced personnel.

Studies showed that the accident at Three Mile Island actually had little effect on human, animal, and plant life in the area. Upgrades in equipment, training, and plant design as a result of the accident greatly improved the safety at nuclear power plants. Nonetheless the public outcry about the potential dangers of meltdown and escaping radiation hobbled nuclear power as an energy source in the United States. Although some plants already under construction were completed and launched after Three Mile Island, no new plants have been ordered since the accident.

Yet another blow to nuclear energy came on April 26, 1986, with the world's worst-ever meltdown at the Chernobyl reactor in Ukraine. A combination of a flawed plant design and inexperienced personnel led to the disaster. A sudden surge of power during a systems test at

No More Nukes and Nuclear Proliferation

In the 1970s and 1980s there was a growing movement worldwide to eliminate nuclear weapons. Protestors by the thousands marched with signs bearing slogans such as No More Nukes and Ban the Bomb. The protestors agreed with Nobel Prize–winning scientist George Wald, who warned, "Nuclear weapons offer us nothing but a balance of terror, and a balance of terror is still terror." The United States and the Soviet Union signed treaties to limit their arsenals and ban certain kinds of nuclear testing. After the collapse of the Soviet Union in 1991, Cold War fears were replaced with apprehensions about smaller, so-called rogue nations obtaining nuclear weapons. In 2004 Abdul Qadeer Khan, the physicist responsible for Pakistan's bomb, revealed that he had been selling crucial information about how to make nuclear weapons to regimes that included North Korea, Libya, and Iran. North Korea has conducted three nuclear tests since 2006. Some fear that Iran will be next to join the nuclear club. President Barack Obama's administration, along with other nations including Great Britain, France, and Germany, has demanded that Iran dismantle large parts of its program to enrich uranium, a step that can produce materials for a bomb. Iran insists it is using uranium only for peaceful purposes and claims the right as a sovereign nation to develop nuclear energy. Nevertheless, Iranian officials are reluctant to allow inspectors from the International Atomic Energy Agency (IAEA) to examine its nuclear sites. Despite economic sanctions, Iran continues to edge closer to having bomb-making capabilities.

Quoted in *Chicago Journalism Review*, "A Generation in Search of a Future," May 1969.

the plant caused an explosion and fire that killed two workers and released dangerous levels of radiation into the atmosphere. In the months following the accident, twenty-eight people died from acute radiation sickness. Large numbers of people in Ukraine, Belarus, and Russia suffered disruptions in their lives due to fears about radiation. The former Soviet Union initially denied reports of the accident, but eventually secretary general Mikhail Gorbachev admitted the facts. Effects from the disaster spread to Western Europe, where thousands of pregnant women were advised to have abortions because of radiation exposure, a measure that experts now believe was unnecessary. The United Nations World Health Organization estimates that four thousand people will eventually have died as a result of radiation exposure linked to Chernobyl, but other groups claim the death toll already is much higher.

Three Mile Island, Chernobyl, and Fukushima have joined the names Hiroshima and Nagasaki as symbols of the frightening power of the atom. Today scientists continue to work on ways to harness nuclear power, including refinements such as cold fusion, which could produce cheap energy from the atoms in seawater. However, many people remain wary of nuclear energy in any form.

CHAPTER FOUR

Atomic Medicine: From X-Rays to DNA

At the end of August 2014 Brett and Naghemeh King whisked their five-year-old son Ashya out of a British hospital and took him to Malaga, Spain. The Kings had grown impatient with their son's doctors and were seeking a more advanced treatment for his brain tumor. They believed that proton beam therapy, a form of nuclear medicine, could save the boy's life. Unlike ordinary radiation treatments, which penetrate through the body and damage healthy tissue, proton beam therapy directs a more accurate flow of particles that stop once they hit their target. However, while proton therapy reduces harmful side effects, research shows that its success rate at treating cancer is about the same as that for radiation. Ashya's doctors had rejected the treatment as unlikely to succeed.

Days later the Kings were released from a Spanish jail after being arrested for moving their gravely ill son. Their arrest outraged those who believed the parents had every right to seek the best treatment. Jeremy Hunt, Great Britain's health secretary, acknowledged that the National Health Service does fund proton beam therapy, but the only proton beam machine in the country is a low-energy device used chiefly for eye surgeries. With regard to the technology, he added, "It's not always appropriate, it's not always safe."[22] Ashya's parents finally took their son to Prague in the Czech Republic for proton therapy treatment, and he is now expected to make a full recovery. The Kings' belief in the treatment is typical of how the public views nuclear medicine today. Atomic theory has led to technologies for diagnosing and treating illness that are used every day and are remarkably effective. Science continues to harness the properties of the atom to save lives.

Roentgen and X-rays

One of the most powerful diagnostic tools in medicine is the X-ray. The existence of X-rays was discovered almost by accident in 1895. Wilhelm Roentgen noted that attempts to photograph electric current passing through a Crookes tube left the photographic plates fogged and overexposed. Ruling out visible light as the cause, Roentgen sheathed his Crookes tube in black cardboard and ran an experiment. When he launched the current in the tube, a specially prepared screen coated with fluorescent paint began to glow, even though the screen was no closer than an adjacent desk. Roentgen found that the radiation from the tube passed through many objects—books, pieces of wood, even his own body. At his request, Roentgen's wife Anna placed her hand between the Crookes tube and a photographic plate. A ghostly image of the bones in her hand appeared on the plate. The effect must have been quite eerie, as Anna reportedly exclaimed, "I have seen my own death!"[23] Roentgen called the uncanny flow of charged electron particles X-rays, a temporary name that stuck.

Roentgen's discovery revolutionized medical science. Diagnoses of internal ailments could now be made without cutting into a patient's body. Because the rays passed through soft tissue but were absorbed by denser matter, X-ray images were able to reveal kidney stones and tumors. A year after Roentgen's work, an X-ray section opened at the Glasgow Royal Infirmary in Scotland—the world's first radiology department. X-rays showed a penny lodged in a child's throat and a needle in a seamstress's hand. During World War I, surgeons employed X-rays to identify cracked bones and embedded bullets. In 1927 a Portuguese physician named Egas Moniz developed a technique of x-raying arteries by injecting them with a contrasting agent. Angiograms, as they are called today, are crucial in monitoring heart disease and detecting blocked arteries. More recent developments in X-ray technology include computed tomography or CT scanning, in which cross-sectional images like slices of certain areas of the body are combined into a three-dimensional image, and fluoroscopy, which is the use of X-rays to provide a study of moving body structures, much like an X-ray movie.

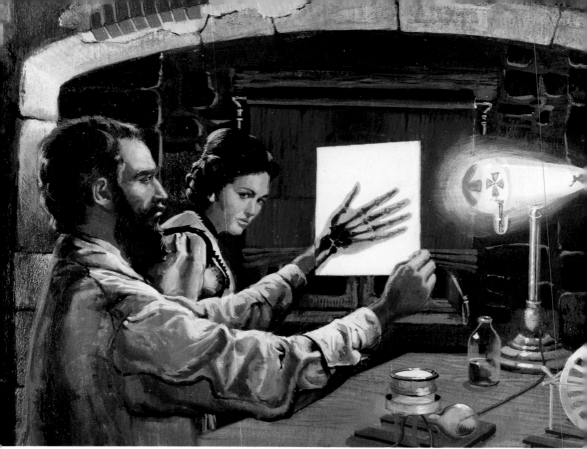

Wilhelm Roentgen's discovery that radiation passes through objects, including the hand of his wife Anna, revolutionized medicine. X-rays, the name Roentgen gave to his discovery, allow doctors to diagnose internal ailments without cutting into a patient's body.

Early researchers working with X-rays soon discovered their dangerous effects. Patients exposed to the radiation for two hours at a time developed itching, irritation, and inflamed skin that resembled severe sunburn. Physicians finally realized that prolonged or repeated exposure can damage tissue and increase risks for cancer. This ability to destroy cells makes X-rays effective in treating malignant tumors. However, public awareness of the danger from X-rays has led to overreaction. Some patients are reluctant to submit to simple procedures such as dental X-rays. In reality, routine chest or dental X-rays pose no safety risk, nor do X-ray machines used in airport security. In any case radiologists and X-ray technicians are trained to employ the least amount of radiation necessary to get the results they seek. Certainly the benefits of this technology far outweigh the costs.

The Manhattan Project and Radiation Research

During the Manhattan Project medical researchers confronted another technology related to atomic theory: nuclear radiation. This phenomenon was still poorly understood. For the thousands of people working with radioactive materials in unprecedented quantities, the potential risks and health effects were unknown. Yet some of the scientists had heard frightening stories. In the 1920s thirteen New Jersey women hired to paint clock dials with radioactive dye—and taught to clean their brushes with their mouths—had died of jawbone cancers. Scientists realized that the effects of radiation might not be apparent for long periods after exposure. They also feared endangering the project by letting the outside community of Los Alamos, New Mexico, become aware of the hazards.

Doctors quietly set up a program called health physics, in which they monitored the health and safety of workers who were dealing with uranium and plutonium on a daily basis. They compiled data from radiation detectors, blood and urine samples, and physical examinations. "The clinical study of the personnel," wrote the project's health director Robert Stone in 1943, "is one vast experiment. Never before has so large a collection of individuals been exposed to so much radiation."[24] Much less defensible were the experiments on civilian patients in Los Alamos and at university hospitals in Berkeley, California; Chicago, Illinois; and Rochester, New York. Subjects were unknowingly injected with plutonium and other radioactive materials to measure the effect on their tissues. Revelations about these tests in the 1990s outraged the public.

> **WORDS IN CONTEXT**
>
> *radiologists*
> Physicians who use radiation-based imaging technology to diagnose and treat disease.

Radioisotopes: Amazing Tracers

Other medical uses of radioactive material also underwent testing. One breakthrough was in radioactive isotopes, which are unstable variants of elements that emit radiation. Scientists had recognized the value

X-rays in the Media of the 1890s

Wilhelm Roentgen's inadvertent discovery of X-rays not only had a huge impact on science and medicine but also quickly captured the public's imagination. While some atom-related discoveries were almost impossible for an ordinary person to understand, X-rays had a simpler appeal: They allowed a glimpse inside people. Before Roentgen's work the only accurate representations of the inside of the human body were anatomical drawings and photographs. Few besides medical students and police surgeons had observed sliced-open cadavers. X-ray photographs thus were novelties that combined curiosity with the macabre. A year after their discovery, X-rays became all the rage, from books to postcards to vaudeville. For two pence Londoners could see their own bones in an X-ray screening at Hyde Park. The 1896 book *Something About X-rays for Everybody* provided a layman's explanation of the phenomenon and even advised hobbyists on how to build their own X-ray machines.

Before long the idea of X-rays was being used for satirical purposes. Newspaper cartoons portrayed X-ray images of politicians, their brains filled with thoughts of booze and thievery. A French postcard depicted an X-ray of a horse-drawn carriage with a skeletal young couple embracing inside. The caption warned the young lady that her escort's intentions were transparent. X-ray photography also was a boon to spiritualists. X-rays seemed to reveal death working beneath the surface of a person's skin. If it was possible to photograph invisible rays, they argued, then it might also be possible to capture ghosts and spirits on film.

of using radioisotopes as a tracer decades earlier. In 1923 a Hungarian chemist named Georg von Hevesy employed a radioisotope of lead to trace how lead moves from soil into bean plants. A few years later, researchers in Boston, Massachusetts, injected patients with dissolved radon as a tracer to measure blood circulation rates.

Radioisotope tracers showed amazing promise for studying the workings of the human body. In the past such research involved extracting and studying dead cells or injecting traceable chemicals.

Radioisotopes were much more effective. Physicist Heinz Haber explains the procedure in a book called *Our Friend the Atom:* "Making a sample of material mildly radioactive is like putting a bell on a sheep. The shepherd traces the whole flock around by the sound of the bell. In the same way it is possible to keep tabs on tracer-atoms with a Geiger counter or any other radiation detector."[25]

One problem was that certain elements in living organisms do not have naturally occurring radioisotopes. This limited the atoms that could be replaced by their radioactive siblings and used as tracers. The solution lay with an invention by University of California physicist Ernest Lawrence. The cyclotron, or particle accelerator, could propel particles to higher energy levels to create radioactive isotopes of various elements. In 1937 Lawrence's brother John, a physician, treated leukemia patients with tracers of radiophosphorus created in the cyclotron. Similar treatments for other blood diseases became standard procedure. In 1939 Ernest Lawrence built a larger cyclotron that was ideal for medical purposes. Doctors began dreaming up new ways to target cancers and other diseases with tailor-made radioisotopes. For example, medical researchers at Massachusetts General Hospital developed radioactive iodine from ordinary iodine, which is essential for proper function of the thyroid gland. The researchers used radioactive iodine to find and treat thyroid cancer, a therapy still used today.

Nuclear reactors built for the Manhattan Project greatly increased the production of radioactive materials. Research on radioisotopes was able to proceed on a larger scale. In the years after World War II, the US Atomic Energy Commission made isotopes available for medical research and therapy as well as industrial use. Today these radiation-emitting atoms are used for everything from diagnosing and treating disease to sterilizing medical equipment.

Other Medical Imaging Technologies

Other technologies based on atomic theory also help doctors diagnose internal ailments. Magnetic resonance imaging (MRI) employs magnetism and radio waves to produce detailed pictures inside the human body. MRI is based on a physics phenomenon first observed in the 1930s called nuclear magnetic resonance.

Atomic nuclei reveal themselves by absorbing or emitting radio waves when they are exposed to a strong magnetic field. In 1970 Raymond Damadian, a physician and researcher in Brooklyn, New York, found a practical use for MRI as a medical tool. "I thought, if we could do on a human what we just did on that test tube, maybe we could build a scanner that would go over the body to hunt down cancer," says Damadian. "It was kind of preposterous. But I had hope."[26] Damadian discovered that the radio waves from cancerous tissue persist longer than those from healthy tissue. This provides the basis for an image, as the MRI is essentially a computerized map of the radio signals a human body emits. The use of radio waves makes MRI safer than technologies such as X-rays or CAT scans, which employ potentially harmful radiation. Today doctors around the world, including neurosurgeons and orthopedic surgeons, perform millions of MRI examinations each year. MRI is vital to the diagnosis of spinal injuries, heart disease, and diseases of the central nervous system such as multiple sclerosis.

> **WORDS IN CONTEXT**
>
> *metabolic processes*
> Organic processes in a cell or organism that are necessary for life.

Positron emission tomography (PET) is a scanning technology used mainly to detect tumors. A PET scanner traces glucose-mimicking molecules that are injected into patients. Since malignant tumors process more glucose than benign tissues do, a PET scan can identify dangerous growths more readily than MRI or CT images. (CT stands for computerized tomography.) The first positron scanner device was invented in 1952, but it was not until the 1970s that PET scanners became a medical tool. In the 1990s Swiss scientists developed a combined PET/CT scanner that could reveal both anatomic details and metabolic processes in the body at the same time. Today's PET technology requires very expensive equipment and highly trained personnel, but its effectiveness as a cancer-fighting tool makes the extra cost worthwhile. Overall, atomic theory has been hugely successful in suggesting a variety of ways to scan the human body for signs of disease.

DNA, the Basic Molecule of Life

Another groundbreaking discovery related to atomic theory is DNA—or more precisely the structure of the DNA molecule. DNA

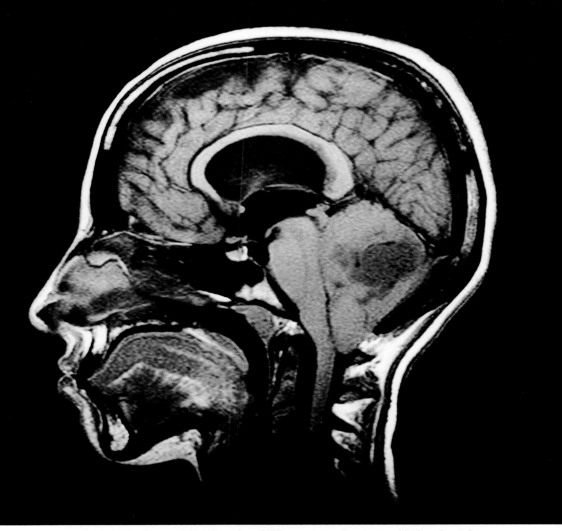

A colored magnetic resonance imaging (MRI) scan reveals a tumor (dark red patch on the right) in the brain of a twenty-three-year-old patient. MRI uses magnetism and radio waves to produce detailed pictures inside the human body for diagnosis of all sorts of injuries and illnesses.

stands for deoxyribonucleic acid. It is a complex molecule that carries the genetic information for nearly every living organism. It is like the instruction booklet for how to build and maintain different life forms.

DNA itself is not a recent discovery. In 1869 Friedrich Miescher, a Swiss chemist, isolated DNA as a nucleic acid while examining various proteins in white blood cells taken from pus-coated bandages. Miescher found that the new substance had chemical properties that were unlike proteins. Through the years scientists continued to

The Promise of Nanomedicine

Nanotechnology deals with manipulating extremely small bits of matter. It takes place on the level of atoms and molecules. One especially promising area for this technology is called nanomedicine. Researchers hope to enable doctors to treat disease and physical injury at the molecular scale. Some even see the potential for extending human life spans.

American physicist Richard Feynman raised the prospect of nanomedicine more than fifty years ago. "A friend of mine . . . suggests a very interesting possibility for relatively small machines," Feynman remarked. "He says that, although it is a very wild idea, it would be interesting in surgery if you could swallow the surgeon. You put the mechanical surgeon inside the blood vessel and it goes into the heart and looks around. It finds out which valve is the faulty [one] and takes a little knife and slices it out. . . . [O]ther small machines might be permanently incorporated in the body to assist some inadequately functioning organs."

Feynman's musings proved prophetic. Researchers today are developing tiny particles that are attracted to diseased cells in the body. The particles can repair or eliminate the bad cells without harming healthy cells. Another type of nanoparticle attacks a virus by delivering an enzyme that prevents the virus's molecules from reproducing. Scientists are also engineering microscopic robots, or nanobots, that can repair chromosomes in cells. With its endless applications, nanomedicine might well become the most important medical technology in the twenty-first century.

Quoted in Mallanagouda Patil et al., "Future Impact of Nanotechnology on Medicine and Dentistry," *Journal of Indian Society of Periodontology*, May–August 2008. www.ncbi.nim.nih.gov.

probe the secrets of DNA. Three years before Miescher's discovery a Central European monk and botanist named Gregor Mendel had described the basic process of genetics, but scientists did not understand the connection between genes and DNA. In 1944 American researcher Oswald Avery proved that genes, which pass on traits from parents to children, are encoded on the DNA molecule. Austrian chemist Erwin Chargaff was astounded by Avery's findings: "This discovery, almost abruptly, appeared to foreshadow a chemistry of heredity and, moreover, made probable the nucleic acid character of the gene. . . . Avery gave us the first text of a new language, or rather he showed us where to look for it. I resolved to search for this text."[27]

Hopes and Fears About Genetic Research

Nine years after Avery's discovery, the English scientists James Watson and Francis Crick described the atomic structure of a DNA molecule as a three-dimensional double helix—like a circular staircase in two sections.

Their work led to a whole new area of science: molecular biology. Researchers began not only to study the processes of life on the subatomic level but to experiment with ways to affect these processes. The Human Genome Project enlisted scientists from around

> **WORDS IN CONTEXT**
>
> *genome*
> All the genetic material of an organism.

the world to collaborate on mapping and understanding human genes in their entirety. Results from this vast project were published in 2003.

A geneticist examines the results of DNA sequencing. Sequencing determines the make-up of genes, which are the sections of DNA that encode how every cell works. Genetic research has provided important new tools for tackling health problems at a molecular level.

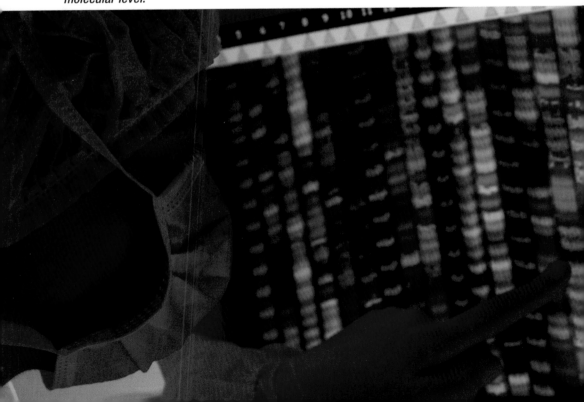

Francis Collins, the director of the National Human Genome Research Institute, says, "It's a history book—a narrative of the journey of our species through time. It's a shop manual, with an incredibly detailed blueprint for building every human cell. And it's a transformative textbook of medicine, with insights that will give health care providers immense new powers to treat, prevent and cure disease."[28] Based on this research, genetic science offers new possibilities for tackling health problems at the molecular level. For example, medical researchers hope to treat incurable illnesses with gene therapy, transplanting healthy genes in place of missing or defective ones. Already biotech companies are preparing gene therapies for commercial use.

Nevertheless, revelations about DNA and genetic research have left some people frightened by the prospect of scientists meddling with genes. They worry about so-called genetic engineering, capable of manufacturing dangerous food plants or fiendish new viruses that could spiral out of control. They fear that information about a person's genetic makeup, including inherited traits for certain diseases, could be used to discriminate against that person in various ways, such as with higher insurance rates or lost job opportunities.

WORDS IN CONTEXT

cloning
The process of creating an exact genetic copy of an organism.

Public anxiety about genetic research is also reflected in popular culture. The real-life cloning of a placid sheep named Dolly—the first successful cloning of a mammal—came three years after its darker equivalent in the film *Jurassic Park* (1993). In the movie a scientist clones a herd of rampaging dinosaurs from blood found in ancient mosquitoes preserved in amber. The science-fiction thriller *Gattaca* (1997) presents a world where genetic engineering has endowed almost everyone with physical perfection. Those whose genes have escaped tampering are considered inferior and denied advancement. Genetic research, like other technologies owed in part to atomic theory, can seem a boon to society at one moment and a potential catastrophe the next.

CHAPTER FIVE

Chemistry and Consumer Electronics

In the twenty-first century the public has come to expect electronic devices to become progressively thinner, sleeker, and more resilient. Imagine, however, that you could fold your cell phone like a piece of cardboard and slip it in your pocket. This is one of the possibilities raised by a material called graphene. It is the strongest, thinnest material in existence. According to the American Chemical Society, graphene is two hundred times stronger than steel, and a single ounce of it could cover an area equal to twenty-eight football fields. It can also conduct heat and electricity better than any known material. Made of a one-atom-thick layer of carbon arranged in a honeycomb lattice pattern, it is the hardest material in the world and also remarkably pliable.

Physicists Andre Geim and Kostya Novoselov discovered graphene in 2004 at the University of Manchester. In their attempts to make a thin layer of graphite (the substance in pencil leads), the pair would use scotch tape to clean the graphite surface. One day instead of throwing the tape away, they examined it under a microscope. They were surprised to see an almost impossibly thin and transparent layer of graphite. As Geim describes it: "My physics intuition, developed over the last thirty years, told me that this material shouldn't exist. And if you had asked 99.9% of scientists around the world they would have said the idea of a 2D [two-dimensional] material was rubbish and that graphene shouldn't exist. And in most cases they would have been right, but in the case of graphite or graphene, and a dozen other materials like it, our intuition was completely wrong. You can reach this limit of one-atom-thin layers."[29]

Geim and Novoselov's work not only earned them the 2010 Nobel Prize in physics but also drew the enthusiastic attention of researchers worldwide. The race is on to find the most imaginative uses for this amazing new material, from bendable touch screens to denser and speedier integrated circuits. The revolution in chemistry and materials ushered in by atomic theory continues to change the world.

Polymers and Plastics

Atomic theory and the discovery of how molecules bond together led to the development of many new materials. Chief among these was a new class of lightweight materials called plastics. Plastics are polymers—large molecules (ten thousand or more atoms) formed of long chains of simple molecules connected by chemical bonds. Before polymers, the only materials that could be molded effectively were pottery clays and glass. While these substances could be shaped into items such as storage containers, they were too heavy and brittle for many purposes. Plain rubber also was inefficient because it tended to lose its shape and could melt in hot weather. In 1839 Charles Goodyear stumbled upon a polymer while trying to make an improved rubber. Goodyear found that mixing sulfur with crude rubber at a high temperature and then cooling the mixture produced a new, harder rubber that could snap back into shape. The process he discovered, called vulcanization, occurs because sulfur forms chemical bonds between the adjacent strands of rubber polymer. The bonds form a meshed link with the polymer strands, enabling them to snap back when stretched. The American public came to know the Goodyear name through the automobile tire business and a blimp that would appear at popular sporting events.

> **WORDS IN CONTEXT**
>
> *polymers*
> Large molecules made up of long chains of simple molecules connected by chemical bonds.

Other polymers followed, including a mixture of nitrocellulose and camphor called celluloid. Discovered in 1870 by an American chemist named John Hyatt, celluloid was used to make everything from ping-pong balls to photographic film. In 1909 another American chemist, Leo Baekeland, created the first synthetic polymer from phenol and formaldehyde. "I was trying to make something really hard," Baekeland recalls, "but then I thought I should make some-

A close-up view reveals the structure of graphene, an atomic-scale honeycomb lattice made of carbon atoms. Graphene is the strongest, thinnest material in existence; it is also remarkably pliable.

thing really soft instead, that could be molded into different shapes. That was how I came up with the first plastic."[30] Baekeland called the substance Bakelite. It could be molded into any shape when heated and then hardened permanently as it cooled under extreme pressure. A perfect substance for manufacturing, Bakelite was durable and inexpensive. It also was an excellent insulator for heat and electricity. In the early decades of the 1900s Bakelite was made into radios, telephones, appliances, car parts, cooking implements, and costume jewelry. As if sensing the endless possibilities for this new material, Baekeland chose for his corporation's logo the symbol for infinity.

Plastics Worthy of Respect

Americans have had a love-hate relationship with plastic for the past half-century. Plastic was once the space-age material that made life cheaper and easier for suburbanites. Households stored food in brightly colored Tupperware, and flaming-pink polyethylene flamingos stood guard in the flowerbeds. Children played with Hula Hoops and Barbie dolls. Then the counterculture of the 1960s brought a backlash against plastic as a symbol of tackiness and phony values. In the 1980s and 1990s, as an avalanche of plastic items ended up in landfills, the material became synonymous with waste and environmental disaster.

Rescuing plastic's reputation would seem to be almost impossible, but scientists at an IBM research lab are making inroads. A recent lab accident led to the discovery of an amazing new family of polymers. Researcher Jeannette Garcia forgot to add an ingredient to an ordinary polymer reaction. To her surprise, the mixture formed a new polymer, an incredibly strong one. "I couldn't even get it out of the flask. I had to break the glass with a hammer," Garcia remembers. "The polymer actually survived the glass breaking, so I hit *it* with the hammer, and *it* didn't seem to break. I thought, 'Hmm, there could be something to this.'" Two versions of the new material, nicknamed Titan and Hydro, also proved to be degradable, one in acid and the other in ordinary water. The idea of such a strong material that can be broken down for recycling has scientists excited about plastic again.

Quoted in Devin Coldewey, "Lab Accident Yields Ultra-Strong Material That Recycles Cleanly," NBC News, May 14, 2014. www.nbcnews.com.

The Plastic Revolution

While Bakelite was a major success, the real plastic revolution developed because of World War II. Before the war the availability of natural substances such as wool, silk, and latex limited the need for new synthetic materials. When the outbreak of war cut off supplies of natural materials, military leaders turned to a wide range of desperately needed synthetics. A synthetic fiber called nylon was used to produce ropes and parachutes. A plastic called polyethylene played a crucial role as insulation for telephone wiring and radar electronics. Neoprene, a synthetic rubber, was molded into vehicle tires, engine parts, and machine components. Gaskets and valves for the atomic bomb were made of polytetrafluoroethylene, or Teflon.

Today the plastics industry touches every area of modern life. Scientists have continued to experiment with the molecular structures of polymers to create plastics for specific purposes. For example, amorphous polymer chains—those with no overriding form or order—are used to make transparent materials, such as Plexiglas, food wrap, and contact lenses. Polymers such as polyethylene can be engineered to be extremely lightweight but very sturdy, as with bicycle helmets and bulletproof body armor. Medical uses for polymers include synthetic rubber catheters, latex gloves, plastic tubing, heart stents, and adhesive bandages. Even parts of artificial hips and knee joints are formed from hard plastic. A team of MIT chemical engineers has used nanotechnology to make an ultrathin polymer coating for artificial joints that helps them adhere to the patient's bones more securely. According to MIT engineering professor Paula Hammond, "This would allow the implant to last much longer, to its natural lifetime, with lower risk of failure or infection."[31]

Despite their versatility and usefulness, polymers do not lack for critics in today's society. Environmentalists note that more than 300 million tons of plastic are produced each year, and much of it is thrown away and does not decompose.

Only about 10 percent of manufactured plastic is recycled. Items such as plastic water bottles, grocery bags, and cups are used once and quickly discarded. "Plastics are very long-lived products that could potentially have service over decades," says Richard Thompson, coauthor of a report on plastics in a British scientific journal. "And yet our main use of these lightweight, inexpensive materials are as single-use items that will go to the garbage dump within a year, where they'll persist for centuries."[32] In addition to their troublesome persistence, plastics also can contain harmful chemicals that are absorbed into the human body. Plastic debris and waste in oceans and waterways can injure or poison marine life and disrupt habitats. As a result many scientists are working on a solution. According to Rolf Haden, a researcher at Arizona State University's Biodesign Institute:

> We are in need of a second plastic revolution. The first one brought us the age of plastics, changing human society and enabling the birth and explosive growth of many industries. But the materials used to make plastics weren't chosen judiciously and we see the adverse consequences in widespread environmental pollution and unnecessary human exposure to harmful substances. Smart plastics of the future will be equally versatile but also non-toxic, biodegradable and made from renewable energy sources.[33]

From Transistors to Microchips

Just as polymers revolutionized modern materials, the semiconductor did the same to modern electronics. And the invention of the semiconductor was based on atomic theory and quantum mechanics. According to quantum theory, electrons sometimes behave as particles and sometimes as waves. Working at Princeton University in the 1930s, physicist Eugene Wigner and his student Frederick Seitz figured out how electron waves could make certain materials conduct or not conduct electricity. That is what a semiconductor is: a material that can act as either a conductor or an insulator. Wigner and Seitz found that the atoms of certain substances are arranged so that electron waves travel through them easily to set up electrical current.

In the 1940s Wigner and Seitz's discovery formed the basis for research by scientists at Bell Labs in New Jersey. William Brattain created a makeshift amplifier from plastic, gold foil, and germanium, a semiconducting material. Much more efficient than bulky vacuum tubes, this first transistor could control a large electric current with only a tiny voltage input. Brattain's colleague, William Shockley, improved the design, and the pair of scientists, along with physicist John Bardeen, eventually received the Nobel Prize in physics for their work. Transistors immediately were used to manufacture transistor radios, which became iconic in the 1960s. However, transistors had an even more important capability than amplification. They also acted as a reliable switch between on and off settings in response to voltage. On or off, yes or no, one or two—this is the logic gate of computer processing and digital electronics. The digital age was beginning to appear on the horizon.

Miniaturization of Electronics

The next big step was making transistors much smaller. In 1958 engineers Jack Kilby and Robert Noyce, working separately at different companies, both managed to fit many transistors on a single chip of semiconducting material, thus creating the integrated circuit or microchip. "What we didn't realize then," recalls Kilby, "was that the integrated circuit would reduce the cost of electronic functions by a factor of a million to one. Nothing had ever done that for anything before."[34]

Kilby and Noyce's breakthrough led to a startling miniaturization of electronic devices that continues today. Kilby's choice of silicon for semiconducting material became the industry standard. Noyce joined engineer Gordon Moore to form the company Intel and develop the first microprocessor, which basically placed all the functions of a computer on a tiny chip. Around 1970 Moore predicted that transistors would be made smaller and smaller, allowing processing power to double every two years. Moore's Law, as it is called, has proved remarkably prophetic. Today transistors have shrunk to the size of a blood cell, and each microchip holds billions of them. Some scientists predict the coming of a transistor the size of an atomic particle.

The result for the public has been an explosion of consumer electronics, including laptop computers, smartphones, tablets, and video game platforms. Inexpensive electronic devices enable a person to link up to the world in new ways. Facebook, Twitter, Instagram, and other so-called social media sites allow people to share information instantaneously. Each succeeding generation becomes more

> **WORDS IN CONTEXT**
>
> *semiconductor*
> A type of solid material that conducts electricity in some conditions but not others.

adept at navigating this new computerized world and expects amazing innovations as a matter of course. However, critics wonder if too many people are tuning out the world around them in favor of their touch screens and toggle switches. Promises of three-dimensional holograms and virtual reality games, powered by ever-present microchips, may only increase many people's isolation and withdrawal from society.

Laser Technology

One technology that challenges the microchip for its importance to the modern world is the laser. This product of atomic theory—and particularly of quantum theory—is the basis for DVD and Blu-Ray discs, bar-code readers, laser eye surgery, and high-speed fiber-optic communications, among many other inventions. Laser science has its roots in Albert Einstein's experiments in 1917 with photons, or particles of light. Einstein affirmed that a passing photon can stimulate an electron in a higher energy state to drop down to a lower energy level, thus emitting a photon with the same wavelength. In 1953 an American engineer named Charles Townes used Einstein's idea of stimulated emission to make a sort of microwave generator. Then he and a colleague hit on the idea of using the same principle to stimulate photons of optical light instead of microwaves. In 1958 Gordon Gould, a doctoral student under Townes, designed an optical laser but failed to patent the invention. (The acronym *laser*, first used by Gould, stands for Light Amplification by the Stimulated Emission of Radiation.) Two years later American physicist Theodore Maiman was able to build a working ruby-light laser. It was another example of a scientific breakthrough occurring almost simultaneously after years of background research. As Townes later reported, "With official publication of Maiman's first laser under way, the Hughes Research Laboratory made the first public announcement to the news media on 7 July 1960. This created quite a stir, with front-page newspaper discussions of possible death rays, but

> **WORDS IN CONTEXT**
>
> *integrated circuit*
> A set of electronic circuits on a single semiconductor chip.

also some skepticism among scientists."[35] The skepticism proved to be misplaced as lasers quickly became practical tools.

A laser begins with a long tube fitted with mirrors at each end. Energy is introduced either by electricity or by flashing light. The energy stimulates electrons in the tube, raising them to higher energy

LED and the Possibilities of Quantum Mechanics

Commonsense views of the natural world governed technology until the beginning of the twentieth century. But then, according to science writer Jonathan Atteberry, a huge change occurred:

> And then came quantum mechanics, the absolutely baffling branch of physics exploring the very smallest types of matter. The study of quantum mechanics led to some truly astounding conclusions. For instance, scientists found that electrons behave both as waves and as particles, and the mere act of observing them changes the way they behave. Revelations like this one simply defied logic, prompting Einstein to declare "the more success the quantum theory has, the sillier it looks."

Silly or not, the quantum version of atomic theory is responsible for some of today's most useful technologies. One example is LED, or light emitting diodes. LED was first invented in the early 1960s, as scientists raced to test the possibilities of semiconductors and lasers. Researchers at Texas Instruments tested the idea of electroluminescence, passing electric current through a semiconducting material, causing the material to emit photons or particles of light. The Jumbotron screen at sports stadiums is a full-color LED display. But LED is simple compared to the latest ideas based on this technology. Scientists now hope to use LED to create entanglement, in which two particles become correlated regardless of how far apart they are. Einstein called this effect "spooky action at a distance." Scientists believe that entangled particles could one day be used for faster communications and computer speeds far beyond today's standard.

Quoted in Jonathan Atteberry, "10 Real-World Applications of Quantum Mechanics," *Discovery*. www.discovery.com.

levels. As they fall back to lower levels, the electrons emit photons of light at precisely the wavelengths of the tube material. The photons bounce up and down inside the cylinder, causing other atoms to emit photons. Eventually the light pulse builds to high energy, releasing a concentrated beam of light.

Later in 1960 Iranian-born physicist Ali Javan and his team made a working gas laser that operated on helium and neon. Unlike Maiman's model, which made brief pulsations of light, Javan's invention produced a steady continuous beam. In 1969 scientists bounced a laser beam off a reflector plate placed on the surface of the moon by *Apollo* astronauts. This allowed for exact measurement of the distance to the moon. By the 1980s laser light streamed through glass fiber optic cable was replacing bulky copper cable for much speedier and more efficient telephone and television transmission. Surveying crews and construction engineers used lasers to make measurements more accurate than ever before. Lasers were also employed in surgery as incredibly precise scalpels of light. Laser use became common for the most delicate eye surgeries. The ability of a laser to cut while at the same time cauterizing tissue with its heat results in surgery that is almost bloodless and much less prone to infection.

> **WORDS IN CONTEXT**
>
> *laser*
> An acronym for Light Amplification by the Stimulated Emission of Radiation.

Today, along with their practical uses, lasers are employed by researchers in particle physics to probe the limits of the physical world. In the public mind, however, lasers are still connected to science fiction tales and death rays. In fact, Hollywood lags behind the genuine technology when it shows a fiendish laser beam approaching a strapped-down heroine. In reality the laser would be invisible—traveling at the speed of light.

Atomic Theory in the Twenty-First Century

From John Dalton's notion of tightly packed little balls to the wildest discoveries of quantum mechanics, atomic theory has been the backbone of modern science and technology. Few people probably think

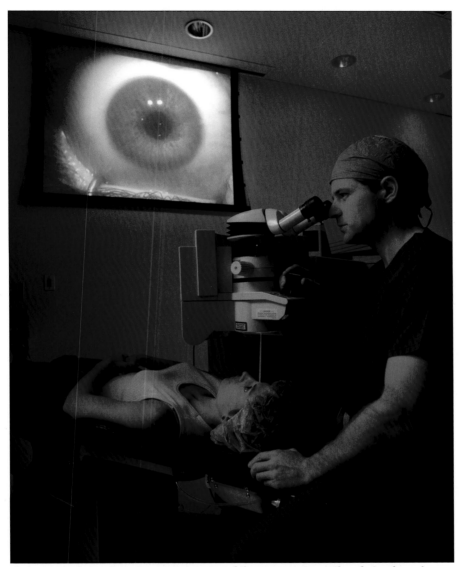

A doctor performs laser eye surgery, one of the many outgrowths of atomic and quantum theory. Laser surgery is almost bloodless and less prone to infection because the laser cauterizes body tissue as it cuts.

about the atomic structure of the chair they sit in or the window they look through, how these objects are mostly made of charged particles and empty space. Yet science's ability to identify and manipulate these miniscule bits of matter continues to shape our lives in unforeseen ways.

SOURCE NOTES

Introduction: A Theory of Tiny Particles with Large Impact

1. Jon Butterworth, "Higgs-Like Discovery from the Inside," *Guardian* (UK), July 4, 2012. www.theguardian.com.
2. Quoted in Adam Hart-Davis, *The Science Book: Big Ideas Simply Explained*. New York: DK, 2014, p. 141.
3. Quoted in the Information Philosopher, "Richard Feynman." www.informationphilosopher.com.

Chapter One: An Ancient Idea Becomes a Modern Theory

4. Quoted in *Hmolpedia: an Encyclopedia of Human Thermodynamics, Human Chemistry, and Human Physics*, "Pierre Gassendi." www.eoht.com.
5. Quoted in William L. Harper, *Isaac Newton's Scientific Method: Turning Data into Evidence About Gravity and Cosmology*. New York: Oxford University Press, 2011, p. 343.
6. Quoted in Marie Boas, *Robert Boyle and Seventeenth-Century Chemistry*. London: Cambridge University Press, 1958, p. 95.
7. Quoted in Ida Freund, *The Study of Chemical Composition*. New York: Cambridge University Press, 2014, p. 290.
8. John Gribbin, *The Scientists: A History of Science Told Through the Lives of Its Greatest Inventors*. New York: Random House, 2002, p. 369.

Chapter Two: The Atomic Theory Develops

9. Quoted in Gribbin, *The Scientists*, p. 399.
10. Quoted in Purdue University College of Science, Division of Chemical Education, "Ernest Rutherford: The Gold Foil Experiment." http://chemed.chem.purdue.edu.
11. Quoted in Claes Johnson, "The Desperation of Planck: The Birth of Modern Physics," The World as Computation, May 29, 2009. http://claesjohnsonmathscience.wordpress.com.

12. Quoted in NNDB: Tracking the Entire World, "Max Planck." www.nndb.com.
13. Quoted in Naomi Pasachoff, *Niels Bohr: Physicist and Humanitarian*. Berkeley Heights, NJ: Enslow, 2003, p. 39.
14. Gribbin, *The Scientists*, p. 515.
15. Quoted in Albert Einstein Site Online, "Albert Einstein Quotes: Religion." www.alberteinsteinsite.com.

Chapter Three: Nuclear Energy and the Atomic Age

16. Quoted in Hiroko Tabuchi, "Reversing Course, Japan Makes Push to Restart Dormant Nuclear Plants," *New York Times*, February 25, 2014. www.nytimes.com.
17. Paul Parsons, *Science in 100 Key Breakthroughs*. Buffalo, NY: Firefly, 2011, p. 283.
18. Quoted in PBS, "Primary Resources: Letter from Albert Einstein to FDR, 8/2/39." www.pbs.org.
19. Quoted in PBS, "American Experience: Edward Teller (1908–2003)," Race for the Superbomb. www.pbs.org.
20. Quoted in Paul Brown, "First Nuclear Power Plant to Close," *Guardian* (UK), March 21, 2003. www.theguardian.com.
21. United States Nuclear Regulatory Commission, *Backgrounder on the Three Mile Island Accident*, U.S.NRC. www. nrc.gov.

Chapter Four: Atomic Medicine: From X-Rays to DNA

22. Quoted in Caroline Davies, "Ashya King's Parents Released from Prison as Arrest Warrant Is Terminated," *Guardian* (UK), September 2, 2014. www.theguardian.com.
23. Quoted in Xfinity, "LIFE.com: Extraordinary X-rays." http://xfinity.comcast.net.
24. Quoted in *Bioethics Archive*, "The Manhattan Project: A New and Secret World of Human Experimentation." https://bioethics archive.georgetown.edu.
25. Quoted in *Bioethics Archive*, "The Miracle of Tracers," Georgetown University. https://bioethicsarchive.georgetown.edu.

26. Quoted in Kasey Wehrum, "How I Did It: Raymond Damadian," *Inc.*, April 2011. www.inc.com.

27. Quoted in Leslie A. Pray, "Discovery of DNA Structure and Function: Watson and Crick," Scitable. www.nature.com.

28. Quoted in National Human Genome Research Institute, "What Was the Human Genome Project?" www.genome.gov.

Chapter Five: Chemistry and Consumer Electronics

29. Quoted in Stefanie Blendis, "Graphene: 'Miracle Material' Will Be in Your Home Sooner than You Think," CNN, October 6, 2013. www.cnn.com.

30. Quoted in Adam Hart-Davis, *The Science Book: Big Ideas Simply Explained*, p. 141.

31. Quoted in Anne Trafton, "New Coating for Hip Implants Could Prevent Premature Failure," *MIT News* (Cambridge, MA), April 19, 2012. http://newsoffice.mit.edu.

32. Quoted in Jessica A. Knoblauch, "Plastic Not-So-Fantastic: How the Versatile Material Harms the Environment and Human Health," *Scientific American*, July 2, 2009. www.scientificamerican.com.

33. Quoted in Richard Harth, "Health and Environment: A Closer Look at Plastics," ScienceDaily, January 23, 2013. www.sciencedaily.com.

34. Quoted in Mary Bellis, "The History of the Integrated Circuit aka Microchip," *About Money*. http://inventors.about.com.

35. Quoted in Laura Garwin and Tim Lincoln, eds., "The First Laser," in *A Century of Nature: Twenty-One Discoveries That Changed Science and the World*. Chicago: University of Chicago Press, 2003, pp. 107–108.

IMPORTANT PEOPLE IN THE HISTORY OF ATOMIC THEORY

Aristotle (384–322 BCE) was a Greek philosopher considered the greatest thinker of the ancient world. He dismissed Democritus's atomism theory. Aristotle's views that all matter is infinitely divisible, that a void cannot exist, and that everything is made up of some combination of earth, air, fire, and water were accepted as dogma by the Catholic Church until the 1600s.

Amedeo Avogadro (1776–1856) was an Italian chemist who contributed to molecular theory with his observation that equal volumes of gases at identical conditions of temperature and pressure have the same numbers of molecules. Avogadro also proposed that simple gases are made up of compound molecules of two or more atoms.

Niels Bohr (1885–1962) was a Danish physicist who applied quantum theory to atomic structure to improve upon the planetary atomic model. Bohr realized that electrons' orbits are actually energy levels and when an electron drops from a higher to a lower energy level it emits energy in the form of a photon.

Robert Boyle (1627–1691) was an Irish chemist who performed important experiments on vacuums, the pressure of gases, and different chemical states. Boyle's insight that an element is something that cannot be broken down by chemical means led to the periodic table and the idea of atomic structure.

Robert Brown (1773–1858) was a Scottish botanist whose observations of the apparently random movements of pollen grains in water, called Brownian motion, led to Albert Einstein's groundbreaking paper on measuring molecules.

Marie Curie (1867–1934) was a Polish physicist who conducted important early research on radioactive materials and coined the term *radioactivity*.

John Dalton (1766–1844) was an English chemist who published the first scientific atomic theory with several important insights. Dalton declared that all matter consists of tiny, indivisible particles called atoms; that atoms of the same element are all alike, while atoms of different elements have different properties; that the whole atom takes part in chemical reactions; and that atoms do not change when they become part of chemical compounds.

Louis de Broglie (1892–1987) discovered that electrons behave both as particles and as waves. This idea, called wave-particle duality, is the basis of quantum mechanics.

Democritus (460–370 BCE) was a Greek philosopher who lived in the fifth century BCE. He supposedly learned about the idea of atoms from his mentor, Leucippus. Democritus's theory of atomism stated that matter was made up of tiny, indivisible particles called atoms.

Albert Einstein (1879–1955) was a German-born physicist who affirmed the existence of atoms with his investigation of a phenomenon called Brownian motion. Einstein also contributed crucial theories about the photoelectric effect and the particle nature of light.

Pierre Gassendi (1592–1655) was a French mathematics professor who revived the ancient Greeks' idea of atomism. He theorized that the shapes of atoms determine their qualities and how they bond together.

Joseph Gay-Lussac (1778–1850) was a French chemist and physicist whose law of combining volumes stated that gases react together in whole number ratios. This observation supported parts of John Dalton's atomic theory.

Murray Gell-Mann (1929–) is an American physicist who discovered the quark, the subatomic particle that is the building block of neutrons and protons, and thus of all matter.

Werner Heisenberg (1901–1976) was a German physicist who used the idea of electrons as particles to perform a mathematical proof of Niels Bohr's atomic model. Heisenberg also framed the uncertainty principle, which states that it is impossible to know simultaneously both the location and speed of an electron.

Antoine Lavoisier (1743–1794) was a French chemist who discovered that air is a mixture of gases and also established the Law of Conservation of Mass. Lavoisier's observations were an important clue about how atoms bond together.

Leucippus (ca. 500 BCE) was a Greek philosopher who died in 370 BCE. Little is known about Leucippus other than his role as founder of the theory of atomism. This theory was developed by his pupil and colleague Democritus.

Dmitri Mendeleev (1834–1907) was a Russian chemist who created the first periodic table of the elements. Mendeleev's chart, arranged by atomic weights and chemical properties, also predicted the existence of elements found later.

Max Planck (1858–1947) was a German theoretical physicist who proposed that radiated energy from atoms is emitted in chunks or packets he called quanta. This idea is the basis for quantum mechanics.

Ernest Rutherford (1871–1937) was a New Zealand-born British physicist who discovered that most of an atom's mass is concentrated in its nucleus. Rutherford conceived the planetary atomic model, with a tiny nucleus orbited by even tinier electrons.

Erwin Schrödinger (1887–1961) was an Austrian physicist who theorized that electrons oscillate around the atom's nucleus in waves. Schrödinger used this wave idea to perform a mathematical proof of Niels Bohr's atomic model.

J. J. Thomson (1856–1940) was an English physicist who discovered the electron and proposed the so-called plum-pudding model of the atom, with negatively charged electrons scattered throughout the atom's positively charged sphere.

Evangelista Torricelli (1608–1647) was an Italian scientist whose invention of the barometer was proof that Aristotle was wrong and that a void, or vacuum, can exist.

FOR FURTHER RESEARCH

Books

Roberta Baxter, *John Dalton and the Development of Atomic Theory.* Greensboro, NC: Morgan Reynolds, 2013.

William Dampier and Margaret Dampier, eds., *Cosmology, Atomic Theory, Evolution: Classic Readings in the Literature of Science.* New York: Dover, 2013.

Theodore Gray, *The Elements: A Visual Exploration of Every Known Atom in the Universe.* New York: Black Dog & Leventhal, 2012.

John Gribbin, *The Scientists: A History of Science Told Through the Lives of Its Greatest Inventors.* New York: Random House, 2002.

Heinz R. Pagels, *The Cosmic Code: Quantum Physics as the Language of Nature.* New York: Dover, 2011.

Anton Zeilinger, *Dance of the Photons: From Einstein to Quantum Teleportation.* New York: Farrar, Straus and Giroux, 2010.

Internet Sources

Kristin Born, "Early Atomic Theory: Dalton, Thomson, Rutherford and Millikan," Education Portal. http://education-portal.com.

Anthony Carpi, "Atomic Theory I: The Early Days," Visionlearning. www.visionlearning.com.

F.G. Gosling, "The Manhattan Project: Making the Atomic Bomb," atomic archive.com. January 1999. www.atomicarchive.com.

Tom Siegfried, "When the Atom Went Quantum," ScienceNews, June 28, 2013. www.sciencenews.org.

M. Mitchell Waldrop, "Alternative Fusion Technologies Heat Up," *Scientific American*, July 24, 2014. www.scientificamerican.com.

Websites

Chemistry I: Atoms and Molecules (www2.estrellamountain.edu). This website features a primer on atoms, molecules, electrons, and chemical bonding. Graphics help the visitor understand the concepts that are discussed.

Intro to Quantum Mechanics (www.quantumintro.com). This website provides a user-friendly guide to the intricacies of quantum theory.

Los Alamos National Laboratory: Periodic Table of Elements (http://periodic.lanl.gov/index.shtml). This website features an interactive version of the periodic table of elements. Visitors can click on an element in the table to find out how it was discovered and what its chemical properties are.

ScienceDaily (www.sciencedaily.com). This website bills itself as "your source for the latest research news." It provides links to articles about the world of science, including material on atomic theory and particle physics.

INDEX